D0996602

# THE WEATHER HANDBOOK

2ND EDITION

ALAN WATTS

# THE WEATHER HANDBOOK

## 2ND EDITION

*WATERLINE*

Copyright © 1999 Alan Watts

First edition published in the UK in 1994
by Waterline Books, an imprint of Airlife Publishing Ltd
Reprinted 1995
**British Library Cataloguing-in-Publication Data**
  A catalogue record for this book
  is available from the British Library

ISBN 1 84037 089 0

The information in this book is true and complete to the best of our knowledge. All
recommendations are made without any guarantee on the part of the Publisher, who also
disclaims any liability incurred in connection with the use of this data or specific details.

Typeset by Servis Filmsetting Ltd, Manchester
Printed in Singapore

# Waterline Books
an imprint of Airlife Publishing Ltd
101 Longden Road, Shrewsbury, SY3 9EB, England

# Acknowledgements

I would like to thank Peter Coles who suggested this title and piloted it into print.

Several of the diagrams are based on ones previously published in books that I have written but which are no longer available.

Fig. 14.3 is after G J Bindon, 'Fog at Liverpool Airport', published in *Meteorological Magazine*, June 1962.

Fig. T.1 is after J S Lawson, 'Caribbean Island Tropical Cyclones 1969–89' published in *Weather*, April 1992.

Photo 3.2 was supplied by the Department of Electrical Engineering and Electronics, University of Dundee.

All the colour cloud studies are from the author's collection.

**By the same author**

*Wind and Sailing Boats*
*Instant Weather Forecasting*
*Instant Wind Forecasting*
*The Wind Pilot*
*Basic Windcraft*
*Cruising Weather*
*Reading the Weather*
*Sailing off the Beach*
*Air Riders Weather*

# Preface

On the wall at the head of our staircase there is one of those decorative frames containing the 'Family Name History' of the Watts family. It tells me that my name is 'of patronymic origin' and means the son of Watt but that Watt stems from the old Germanic 'Walther' and that in its turn meant 'ruler of a people'. We are also told that in the Middle Ages 'the bearer of the name would have commanded great respect.'

Now great respect is not something that I have ever commanded. If I've ever got it then it has not sat easily on my shoulders because I have never wanted to be in any way revered for myself but only for what I have learned to do. As I look back, I see that whenever it looked as if I might be asked to take on a position in society where I might have to compromise with the truth I have chosen to duck away and let others take on the challenge. Some of this stems from my birth under Pisces, but with Aquarius rising. Here lies my Achilles heel. I have always been deep down a volatile artist who has had to earn his crust through the science based squarely on the truth that is physics which, to quote my old maths lecturer 'Sammy' Haskell, is simply mathematics complicated by apparatus. I was always a great disappointment to men like Haskell because while I knew I ought to like advanced mathematics I could never square it with my artistic temperament. I just scraped my pass Degree and then only because I found some bits of the paper where there were patterns to be followed. I suppose I wanted maths to be simply geometry or at least I wanted it to appeal to the artist in me.

Thus when I eventually had to find employment, after I'd gone back to school following serving in the navy for two years of the war as a rating and as an officer for two years after that, I went for a job which was half art and half science. I became a weather forecaster.

Here the dormant artist can assuage his desires by drawing great sweeping isobars. He can fit them together and make them look 'right'. For when isobars look right then very often they are right. This is because nature likes to do things in a way which is deeply satisfying to the aesthetic inner self. When we contemplate the grandeur of nature we are awed but when you become a meteorologist, and take weather apart to try to understand its grand design, it is then that you realise that we still need the intervention of a 'God'. Only in that way can you explain the why of a snowflake – the pearls of altocumulus clouds rippled across the sky and painted in the subtlest of shades by the sunrise. It has to be because it is pleasing to us – a gift sent by someone who loves to see us aglow with wonder.

I well remember the first time I was really deeply stirred by weather. I must have been eight or nine I suppose. This particular Monday I trudged home through the puddles playing 'boats' with matchsticks in the streaming gutters because there had just been one of those great springtime outbursts of rain. Effusions that set the young heart burning with anticipation of the wondrous things that might happen in the future. You are all sunshine and showers when you're young.

I remember how when I got home my mother was hanging out the clothes now that the shower had gone. The joyous gurglings of a starling on the chimneypot made me look up and behold the roof of the house framed by the incredibly white sunlit alabaster mouldings of the cloud which had so recently deluged us.

Here was the promise of life. The brightly washed sunshine, the starling's spring song and the tremendous explosion of nature towering into the sky. Ever since that magic moment the wonder of the weather has never left me.

I was very fortunate as a forecaster. It seemed to be my forté, and fate decreed that I should spend the first couple of years on my own doorstep, Thorney Island air base. This was a place I knew intimately. I had ranged over its salt marshes as a boy, built square-rigged model ships to sail on its dykes and watched the lights of its aircraft doing 'circuits and bumps' as I lay awake in my little back room at home before the war. Now I was somewhere I had always dreamed of being – in the middle of the flying, talking to pilots and crews – trying to satisfy the Wing Commander 'Flying' as to the impossibility of a navigation exercise tonight.

At the same time I was able to assuage another great desire – to own a high performance sailing dinghy and pit myself against the sailing aces who populated the clubs of Chichester Harbour in the middle of which Thorney stood. I wanted a Firefly – the 12-foot one-design by Uffa Fox which was the dinghy on which many of the great post-war racing names cut their teeth. It was being built by Fairey Marine at Hamble near Southampton and one of their executives was Charles Curry. In those days the Olympic Single-handed trials were sailed in Fireflies because the Finn was very new and Charles sealed his place at the Helsinki games of 1952 in *Sunlight*. Charles' wife Bobbie was in the business of selling second-hand Fireflies so I went to her with my saved up pounds and saw what I couldn't afford but what I had to have – *Sunlight*! Much against my mother's will I borrowed the money from an aunt and had the prize of my life.

On interminable night duties I did some research into a means of forecasting when the sea-breeze might come in. On a coastal airfield the wind from the sea often came suddenly, reversing the previous wind and throwing the landing of fairly low-powered aircraft into confusion. Knowing when the breeze would arrive was important. However it was even more important to one who sought to win sailing races. If I knew when the wind might shift by a very large amount, think of the races I might win.

It was now that I started a long association with *Yachting World* and its editor Bernard Hayman. I wrote up the gist of the new-found knowledge of the sea-breeze and how it worked, got my fiancée, who is now my wife, to type it, and sent it in. It was accepted and I went on to write a long string of articles for Bernard. Out of that experience I found I had enough material for a book. I well remember the heat of excitement when I came out of a call box with the news that Adlard Coles Ltd had accepted *Wind and Sailing Boats*. The first book I had ever written was going to be published! I was staggered by the favourable comments from American critics when it was published over there and even now the book will not entirely lie down.

Flushed with this success I went on to write a string of other books on weather for Joe Public both ashore and afloat. In particular I remember being badgered by

James Moore, who was the Adlard Coles Ltd editor at the time, during a very hot summer to finish a book with twenty-four colour cloud pictures and associated explanations of what they meant. I didn't want to do that – my scientific self said that the text-book of physics I was writing was what I should be doing. The text-book never saw the light of day but *Instant Weather Forecasting* got into the 'Hidden Best Sellers' list, has been solidly in print for the last thirty years, has survived three take-overs and has seen ten different language editions. That's how stupid you can be!

I have, since then, learned a few things and I am assured by my editor Peter Coles, that it is because of these books and articles that have gone all over the world that I am permitted to write a book for him called *The Weather Handbook*. I just hope that you will find the experience as rewarding as I have in putting down the knowledge acquired over some sixty years. That is the time span if the experience of an eight-year-old can be assumed to have started a meteorological career.

# Contents

Chapter One

# Significant Weather

I live in a meteorological no-man's-land called north-east Essex. We lie under that fuzzy line of division between the forecast areas of East Anglia, which is further north, and south-east England which is further south and really the other side of the Thames Estuary. When we get the forecast for East Anglia we always have to ask 'Do they mean us or should we be under the south-east England bit?'

There are thousands of people like us all over the world, who for one reason or another often do not think that the forecast applies to them. Sometimes this is because of the peculiar pattern of hills and valleys, rivers and coastlines that lie upwind of them. Sometimes it is because of the impossibility of being able to give sufficiently detailed descriptions of the weather in the very limited time that the media allow for each forecast slot. When you sum up the total time devoted to forecasts by all the radio and TV stations of your area it amounts to a sizeable period but as they are all saying the same thing in slightly different ways it just means that a fuzzy view of the weather gets reiterated time after time.

This state of affairs stems from the remarkable revolution that has occurred in the business of forecasting by computer. There is no doubt about it – and you can accept it from me, an ex-met man – the forecasts that come from a happy marriage of human forecaster and

machine are much more accurate and detailed than they were in the old days when we had to do it by hand. Now the most massive computers in the world crunch immense sets of numbers and regurgitate reams of charts of what the isobars and upper-air contours are going to look like tomorrow, the next day and maybe six days ahead. Together with satellite cloud pictures taken both in visible light and in infra-red, radar plots of where the rain is falling and an onward-going better appreciation of the processes of weather, the computer/human method of forecasting is a great success.

I was a forecaster back in the fifties but I have tried not to lose contact with the amazing growth of the forecasting game since then. However that does not often stop me being amazed when they get the general weather right for say Friday, when it is as yet only Monday. We could never have done that in our day except in exceptional circumstances. Today it is done every day, day-in and day-out on 365 days a year with an extra day when it's leap year.

In my day there were many more airfields etc sending in observations than there are today. Weather forecasting starts with observations. If it were not for the unsung heroes whose job it is to observe the weather across the entire world, and sometimes in the most remote and inhospitable places, there would be no forecasts. Computers are wonderful but just like the loveable robot Johnny Five in the film *Short Circuit* they need input.

So how have the meteorological powers-that-be bridged the immense gaps that have appeared in the observing network? The answer is, with the computer which works with the so-called equations of motion of the atmosphere and can fill in the spaces. It could be more accurate if there were more observing stations but there aren't and the number could get fewer.

Even so, the computer has its limitations. It does not know about clouds and rain and fog. All it knows how to do is draw isobars. It cannot even find fronts. All that has to be done by the human forecaster so it is not surprising that while the general trend of the weather chart is pretty accurately portrayed by the machine, the actual weather that clothes those raw isobars has to be divined by the met man on the bench. Little wonder that there is still room for the would-be lay forecaster to refine the official story and maybe get it more accurate than any of them for their own patch.

On one particular day I recall that thunder was in the air. The morning forecasts promised thunderstorms for our region. The wind, from my wind vane in the garden, was south and light. You will find in the following chapters that I believe that those who would foretell the weather must take the trouble to rig up a reliable weather vane. If they cannot then they must expect that their score in local forecasting accuracy will be unfortunately low.

All the morning nothing happened. The lunchtime forecast told us to expect heavy storms to arrive in parts of our forecast area but there was no real sign in the sky of impending trouble. Why the delay and why did it matter?

There are a thousand and one reasons why people hang on the weather forecast. They may want to hang out washing, they may want to go to the beach, they may just want to know if they should encumber themselves with a raincoat or an umbrella. We wanted to know because we had horses out and half an hour's drive away and we'd just scared ourselves by reading of cases where horses had been struck by lightning. We needed to get them in but not unnecessarily early.

There were all the ingredients for massive storms. The relative humidity should by rights be lowest around the middle of the day. This day it was 80% at lunchtime and a Fahrenheit temperature to match.

As we anxiously watched the weather that afternoon the wind fitfully increased and the clouds gathered. We desperately needed to know if there were storms about. So we listened for atmospherics on the radio but could not detect any. Then all the afternoon's cloud passed without incident. It was a bit of a let down. Where were the promised thunderstorms?

When we got back from seeing the horses safely in their stables the wind had backed to ESE and there were thundery-looking clouds coming in from SSW. We shall see that these wind directions which are ninety degrees to one another are a potent sign of a developing stormy situation. So, despite the hot sunshine that persisted into the evening we were not surprised when the evening forecast went for very heavy storms in the West, moving our way overnight.

It began to thunder at about 21.00 and when the storm fully broke the lightning was tropical in its almost continuous intensity. At one stage there was a linked chain of lightning that ran along the clouds to the south and ending in a sort of convoluted knot-like loop. Another chain ended in a ball which disappeared as fast as it appeared. It was a strange storm for it came in waves. Each time a wave arrived the wind swung west and then reverted to east once it had passed on.

We were lucky. The major part of the storm seemed fixed to the south of us. A neighbour had been to an evening event down there and reported that the sky was an almost continuous blue light, so bright that he hardly needed the car's lights. At a small town to the north of us cars were floating down the high street. To the south of us an inch of rain fell in an hour and in my experience an unheard of event occurred. A car was struck by lightning and exploded leading to a bomb scare. Cars usually seem to be particularly immune and even now I would say to anyone who is scared of being struck, 'Get into the car, you'll be safe there'.

From all over the region there were tales of houses struck by lightning and accidents on the motorways. But by far the commonest trouble was households that were flooded out by the torrential rain. Not many miles away, while we personally escaped with hardly any rain to speak of, a great swathe of the coast was washed out by nearly two inches of rain.

This experience has some lessons for us. From the forecast point of view, the weather promised turned up, but much later than expected. The morning forecast had however emphasised that this was going to be no ordinary storm. It is often not the weather but its timing that the forecasters get wrong, and here is where people on the spot can do a great deal for themselves. The message of this book is going to be '*Listen to the forecasts, note what they say and then use your own knowledge to refine the details for your own area.*'

The other aspect to notice is how incredibly variable weather can be within a relatively short distance. Six miles is in weather terms no distance at all and yet it can make all the difference between, in this case, the brilliant but not threatening storm we experienced or a truly terrifying ordeal made doubly bad by local flooding. Such storms only turn up every so often thank goodness and it will be explained in the chapter on thunderstorms why this is so.

Significant weather, as far as the met services are concerned, is weather that really affects the people at whom it is directed. For jet airliners plying the Atlantic air-routes it is the high-speed upper winds and thunderclouds, some of which are lost in other cloud masses and so become a hidden danger. For the public it is rain and showers, strong winds, hail and thunder, snow and icy roads – many more weather elements affect the earth-bound dweller. For the para and hang-glider fraternity, as well as balloonists, it is whether the wind will be above 15 knots at their altitude and if there is any convective clouds with updraughts and lift-destroying rain. For the small-craft sailor it is how strong the wind will be and its direction. Whether there will be poor visibility or good visibility and for the dinghy or board sailor, any aspects of weather which will in any way affect the strength and direction of the wind in small waters.

All these different interests cannot be addressed by the media coverage of weather and if you need more detailed information then telephone forecasts specifically directed at your needs are the best bet. The small-craft sailor may listen to shipping forecasts but be unhappy with what he is told because they are, as they say, for shipping who do not much worry unless the wind gets to 30 knots or so and the visibility drops to fog limits. They also do not really cover that zone of variability – the coast –

where most small-craft sailors sail. You can get forecasts and charts by fax but you have to pay for them. You can connect up a suitable side-band radio to your computer and receive radio weather fax or buy a tailor-made weather fax machine and be able to programme the device to churn out the charts you find most useful at the times you set on it. If you want instant charts, video the main weather forecast presentations and freeze the charts they show so that you can study them. They are the up-to-the-minute thinking on what today's and tomorrow's chart will look like. They sometimes give you continuity charts showing how a low is expected to move across your area.

Thus there is a plethora of weather information to be had and you might say that makes this book redundant. I do not believe that or else I would not have written it. For one thing, knowledge of weather systems helps interpret the information given through the media and, as pointed out before, sometimes the forecast weather is delayed or maybe does not come to you at all. There is still room for the lay person in the forecasting game.

Chapter Two
# Looking at the Sky

Most people, when they look at the sky, view it as a painting looked at from afar. Very often they do not see the individual clouds but rather they take in the sky's visage as they might do a friend's. Is the sky smiling or is it scowling? Is it peaceful or is it angry?

That is perfectly right if you are looking at the sky as an object of beauty or interest. It is not going to get us very far if we wish to look at the sky to foretell the weather.

To make sense out of a skyscape you often have to do more than get an overall feeling for its meaning. Rather, you must do what an artist would have done if he'd painted the sky. Take it apart and see what there is and which clouds are the most important and which is in front of which. In other words split the sky's clouds into decks and then identify the clouds which can mean something and ignore the others.

One of the things that puts many people off trying to foretell their own weather is the prospect of the enormity of the task. The weather has so many aspects and moods that no one is going to be able to hoist it all in. After all, the professionals spend a lifetime trying to do it and they can still get it wrong, so what chance for the person in the street?

There is only one way round this problem. Take a limited objective. Not

every aspect of weather is a threat so concentrate on those weather systems that are, or may be. That is what we are going to do in this book. Whether we like it or not it is threatening weather, be it rain, snow, fog or ice that we must know about. Fair skies deserve consideration but it is skies which foretell coming trouble which are vitally important.

**Cloud decks**

The cloudy part of the atmosphere lies below an invisible 'lid' called the tropopause. Clouds cannot penetrate this top that is put on the weather and about the only time when we actually can see its effect is when thunderclouds spread their anvils under it. The layer below the *tropopause* is called the troposphere and all weather processes of interest to us occur in the *troposphere*.

The clouds of the troposphere split themselves rather neatly into three height layers or decks. (Fig. 2.1). The highest are feathery ice clouds above about 20,000 feet (6km). All these clouds have names that start '*cirr-*'. The lowest are water clouds below about 7000 feet (2km) and they have no prefix. In between are the medium-level clouds which are all prefixed '*alto-*'.

There are two major shapes of clouds. They are either heap clouds (*cumulus*) or

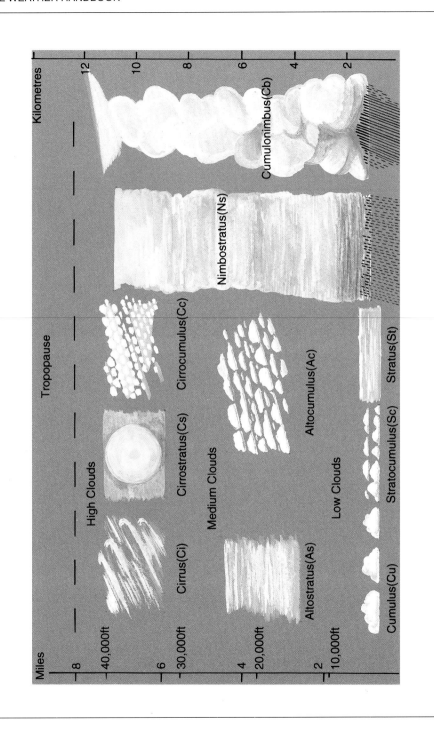

⇐ *Fig 2.1*
*The major cloud types. The heights indicated are very rough and are lower in winter than in summer. However, for our purposes only cloud type is important. The rain-bearing clouds do not always stretch up to the tropopause because of inversions at lower levels.*

layer (stratus), thus we can make up most of the names of the cloud types shown in Fig 2.1.

In the low deck we have cloud names with no prefixes i.e. *cumulus* (Cu), *stratus* (St) and a merger of the two, *stratocumulus* (Sc).

In the medium deck we have *altocumulus* (Ac) and *altostratus* (As). There are also important forms of these two basic types connected mainly with thunder which we will need to recognise.

In the high deck we have a very important cloud, *cirrus* (Ci), another important cloud, *cirrostratus* (Cs) and a not very important cloud *cirrocumulus* (Cc).

If these were the only clouds we would get very little rain because only deep clouds rain at all heavily. The two cloud types which produce most of the precipitation (i.e. anything that falls from clouds) span all the three height decks.

The deep dark cloud type which produce most of our continuous rain is called *nimbostratus* (Ns) while most showery rain comes from the eruption of *cumulonimbus* (Cb). The word nimbus means 'rain-bearing'.

These few cloud types are all we need to get to grips with looking at the sky and using it to foretell the weather.

## The cloud catalogue

*Cumulus* (Cu) – is the cloud of fair weather. To be 'fair weather Cu' it must not be very deep. Look at the cloud in the right foreground, it is not as deep as the base is from the sea. This is a rule of thumb to distinguish a sky which will remain fair and one which might conceivably shower on you later.

Each cloud element is in the shape of a cauliflower because of the way thermals rise into the base and curl over down the sides rather like a fountain might do. Cumulus over or from the sea is less well formed than Cu over or from the land.

*Photo 2.1  Cumulus (Cu)*

Often in the latter case the bases are all at the same level and the tops are usually all about the same height as well. The wind should only be moderate or less.

*Stratus* (St) – is fog above the ground and often not very far above either. Thus it is a great hazard to hill walkers and mountain-climbers and also to road-users who have to venture into high places. It is a cloud particularly associated with the SW wind after warm fronts have passed. When it forms over coastal regions, it does so overnight on the back of moisture-laden winds. Even over low ground the base may not be more than 100–200 feet (30–60m) up. Over-night fog may lift into stratus as the wind gets up in the morning.

*Photo 2.2 Stratus (St)*

*Stratocumulus* (Sc) – is cumulus in a layer and is a cloud of situations which are not changing appreciably. Much of our cloudiness stems from Sc layers. Many old fronts are nothing more than thick banks of stratocumulus and the cloud in the picture may well be an example of this.

Lands which lie in the path of winds from the sea will often not see the sun even when the charts show that an anticyclone covers the region. In the depths of winter extensive layers of dark, dense

*Photo 2.3  Stratocumulus (Sc)*

Sc settle over the centres of static anticyclones leading to days of 'anticyclonic gloom' with poor air quality. At other times Sc is a benign type of cloud often seen in globules or waves with chinks between the elements.

*Photo 2.4 Altostratus (As)*

*Altostratus* (As) – is the layer cloud of the medium levels which is associated with coming rain. While altostratus itself does not rain, it immediately precedes the nimbostratus that does. It is the cloud which gives a 'watery sun' and is often grey and featureless, looking like inverted mud-flats. That is when it is going to rain soon. At other times it just provides extensive islands of dark cloud about the sky. Such sheets of As are often the 'bones' of old depressions and fronts but the weather associated with them has generally died out.

Sometimes it becomes unstable and produces showers which can be thundery in nature when the rest of the weather regime is ripe for it.

*Altocumulus* (Ac) – consists of small globules spread in rafts and islands about the sky. It can take many forms and two important ones will be discussed in Chapter 15. Altocumulus is often found where frontal clouds are dispersing or

*Photo 2.5  Altocumulus (Ac)*

maybe where they are developing. It does not become important until it is accompanied by altostratus.

*Cirrus* (Ci) – seen here at sunset, is feathery and diffuse around the edges because it is an ice-crystal cloud. Obviously ice-clouds are going to be the highest in the sky and so the fibrous appearance of cirrus gives away its height. However the shape of cirrus elements is also unlike any other cloud. In the top centre we see a dense 'head' of cirrus with a 'tail' sinking down towards the right. The tails are called 'fallstreaks' and to the left of the picture we see a whole line of cirrus heads with fallstreaks which are combining together to make a layer. When recognising Ci we need to remember to look for the individual cirrus clouds because so often the most obvious things are the layers, or banners, which might appear to be some other cloud type. However once again there is an important recognition point that will help here. Because they are ice crystals (just as is snow), so Ci clouds are white and in the sun show no dark shadows. The picture shows that they may do so at sunset or dawn.

*Photo 2.6  Cirrus (Ci)*

*Cirrostratus* (Cs) – is a milky veil of ice-crystal cloud that spreads across the sky usually following the appearance of Ci. It is easily recognised because it is the only cloud type to produce a ring halo. There are many forms of halo but the one most often seen is the one shown. A ring, whose radius is about the same as that of the outstretched hand at arm's length, and is only very vaguely coloured, forms about the sun or the moon. If you have the ring halo then you have cirrostratus.

*Photo 2.7  Cirrostratus (Cs)*

23

*Photo 2.8  Cirrocumulus (Cc)*

*Cirrocumulus* (Cc) – is a very rare cloud and even this brilliant display may not be true cirrocumulus. However the picture shows what to look for i.e. small, white globular patches arranged in rafts. It may often be impossible to tell whether the cloudlets are actually Cc or small Ac and from our point of view it is immaterial. Neither cloud has an important prognostic value. They are pretty but inconsequential. This picture illustrates a way in which you can see which clouds are above which. At sunset, the shadow of the edge of the Earth falls on the lower clouds leaving higher clouds in sunshine. We see this here where the As in the background provides a darker contrast to the 'mackerel sky' above.

*Nimbostratus* (Ns) – is the dark, deep rain-bearing cloud of bad weather. Only if you take off from an airport in rain and climb up through the murk into the sunshine above can you realise how deep and often solid nimbostratus can be. Under the overcast it is difficult for anyone to make up their minds whether the cloud they have is Ns or some other lesser layer cloud like St or Sc. However if the cloud rains more-or-less continually, then you are safe to say that you have nimbostratus. From a practical point of view, when gathering As begins to rain, simply change the name to nimbostratus. Nimbostratus is the cloud of fronts and depression centres and it contributes most of the winter rains as well as quite a proportion of the summer ones.

*Photo 2.9 Nimbostratus (Ns)*

*Photo 2.10 Cumulonimbus (Cb)*

Cumulonimbus (Cb) – is the cloud which produces showers and thunderstorms. Each Cb cell is an over-grown cumulus but often masses of individual cells combine together to produce what might look like some very nasty nimbostratus. However, the giveaway is that with Cb the rain starts heavily and tends to tail away before the next deluge. With Ns the rain is not truly showery in nature. It may stop and start somewhat, or some heavier patches come along mixed in with lighter rain, but it is not like showers.

In the picture we have a thunderstorm-cell over the sea with its head pressed up against the tropopause and false cirrus around its anvil top. Surrounding it at low level there is an entourage of Cu clouds which, when it is close, may often screen one's view of the true monster behind. Half-way down the cloud mass we see downward projecting lumps resembling cows' udders. This is characteristic of big Cb clouds and they are technically called 'mamma'. In the United States the sight of mamma is taken as a sign of possible tornado formation. In Atlantic Europe, while tornadoes may occur, mamma more usually means that a mass of Cb clouds is moving away.

Not all Cb clouds will be this massive. In winter and spring snow showers will occur from much smaller cumulonimbus clouds and then it may be difficult to tell that the snow, which seems almost continuous, is not from a front. Even at other times of the year showers can fall from Cb clouds which are much less massive than the one shown here.

If you are new to cloud recognition then these ten cloud types are the ones to start on. However there are other important ones and then there are the cloud-scapes. By that I mean those skies in which there may be several different cloud types at once. These are often the most prognostic of them all and we shall cover them in more detail in the coming chapters.

Chapter Three
# Clouds and Cloudiness

Apart from Venus, ours is the only planet which has clouds. Because of this, Earth from space is a beautiful place – a blue planet speckled and swirled with the delicate tracery of its clouds. In Photo 3.1 we see the view of the astronauts in *Gemini 12* as in one of their orbits they looked eastwards from Mexico over Baja, California. The white speckled areas are cumulus and cumulonimbus with, in the left background, the clouds of an old depression.

It is of considerable interest to realise that the white edge to the Earth shows the relative depth of the atmosphere and indicates how thin is the shell of air we inhabit. Also, these early shots of Earth show it as it really appears to those in space. Later pictures have been 'doctored' to exclude the pre-dominant blue, but the blueness is as Earth is to the naked eye, just as the sky is blue and for the same reason.

There are two types of weather satellite system. The polar orbiting *TIROS* (television, infra-red, orbiting, satellite)

satellites rotate (as shown in Fig 3.1) in orbits over the poles. These provide detailed views of the topography and the clouds both in visible and infra-red light. The infra-red capability allows clouds to be monitored at night and each region gets two passes a day. Photo 3.2 was taken by one of the modern series of *TIROS* satellites and the image has been electronically enhanced.

The other type of satellite system is the geostationary. Six satellites are strung out

*Photo 3.1 The view of the astronauts in Gemini 12 looking eastwards over Baja, California.*

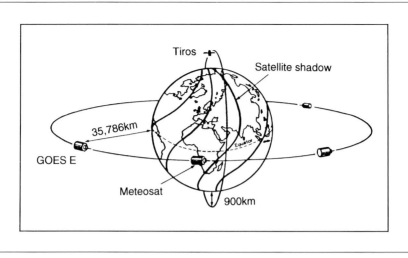

Fig 3.1 The two kinds of orbit used by
meteorological satellites. The
geostationary satellites are almost
36,000 km above Earth while the polar
orbiting TIROS satellites are much lower
and so give much better detail as in
Photo 3.2.

over the equator at a height of 36,000
miles at which height they naturally orbit
at the same angular speed as the Earth.
Thus they remain over one spot and they
scan the clouds of the whole Earth from
their standpoint once every half an hour.
It is these pictures which appear on TV
weathercasts and, when strung together
like the frames of a film, show the movement of the cloud systems. The geostationary satellite for Europe is Meteosat
and stands above the Greenwich meridian. There are two for America GOES
East and GOES West (Geo-Orbitting
Environmental Satellite).

**Satellite cloudscapes**
Clouds from space, where the sun always
shines, are going to be predominantly
white. In modern TV satellite metcasts

this whiteness has been doctored to
make it accord with our earth-bound
experience which says that bad weather
clouds are dark, not brilliantly white.
Thus the clouds above the areas of worst
surface weather are now made to appear
dark.

The brilliant whiteness seen by the
satellites' radiometers is because the
highest layers of the cloud shield over a
depression are cirriform. Ice clouds will
obviously reflect like a snow surface but
they are often over the deepest masses of
nimbostratus. On the ground it is deeply
overcast, raining, snowing, blowing a gale
but the cloud tops are as white as the
driven snow – which in effect they are.

The same goes for the bright white
lumps seen across southern England and
north-west France in the spring satellite
picture (Photo 3.2). These may look
attractive but they represent cold wintery
showers as each lump is a mass of
cumulonimbus cloud. These are
probably the result of the demise of the
cold front whose remains sweep down
across the North Sea from the old
depression between Scotland and

Norway. It may have disappeared as a connected swathe of cloud but its unstable influence is still there just waiting for the spring sunshine to make it erupt into showers.

Many more such nasty weather clouds run down from the northern seas towards Ireland in the wake of the depression. However over Ireland itself the clouds are cumulus for the most part with some bigger ones which could well be producing showers. Over northern France and the Low Countries there are streets

*Photo 3.2 When it is spring the sea is cold but the land is warming rapidly. This makes the land stand out starkly against the sea on satellite pictures. The remains of a cold front trail back from the occluded depression between Scotland and Norway. This old front has aided the eruption of heavy wintery showers over the southern half of England and Brittany. Over northern France fleets of cumulus form along the wind inland but there are bigger shower clouds along the coasts.*

of smaller cumulus but they are more difficult to detect.

**Lows and fronts from space**
The great swirl of cloud over the northern North Sea is just how we might expect a depression (or low) to look from its dartboard kind of representation on a weather map. However it is only when a low becomes middle-aged that it exhibits this spiral form. What we are seeing are again the highest clouds that are travelling in circulating upper winds that are more-or-less parallel to the winds near the surface. We shall see in the next chapter that such a set-up indicates that the system is no longer developing and in fact is in its death-throes. So when a depression actually looks like a depression it is well past its prime.

What does a young active depression look like on the satellite images? When it is first born it may hardly be detectable. Just a bulge on the long swathe of a cold front trailing back from a mother depression. The embryonic low is called a wave depression and we see one in Fig 3.2 plus many other interesting things.

As well as indicating where the fronts are, this composite figure shows which kinds of clouds were to be found and the kind of weather that was underneath them. However the wave can be seen as a bulge where the cold front has become kinked into a wave-like form. The cloud there is only altostratus (As) at present but could well become nimbostratus very soon. There is another wave up towards Iceland and something you do

not often see on TV weather maps – a double frontal system.

These small waves usually run along the cold front quite quickly and sometimes they produce worsening weather for a while and then they die out. At other times they develop into full-blown depressions in their own right.

The 'mother' depression has a signature which met men recognise on the cloud images as a comma cloud pattern. The cloud shield over the centre of the low makes the dot of a large comma while the cold front makes its tail.

So on a satellite cloud image a new depression is an almost imperceptible bulge on a cold front; a maturing depression is a big comma and an old depression is a spiral whirl. The clouds over Ireland and down across Biscay look greyer than the rest because they are stratocumulus (Sc) and so are 'warm' clouds. The radiometer of the satellite sees the coldest parts as white and the warmest as dark grey or black. So as the cirriform clouds over the depressions and fronts may be at something like –40°C, the sea surfaces will appear dark being much warmer and so will the land. The tops of low clouds may be quite cold by our standards but they are warm compared to the cirrus clouds and so they appear grey.

**Holes in the clouds**
There is much to be learned from looking at, and trying to interpret, these satellite images. They give us an overall view and

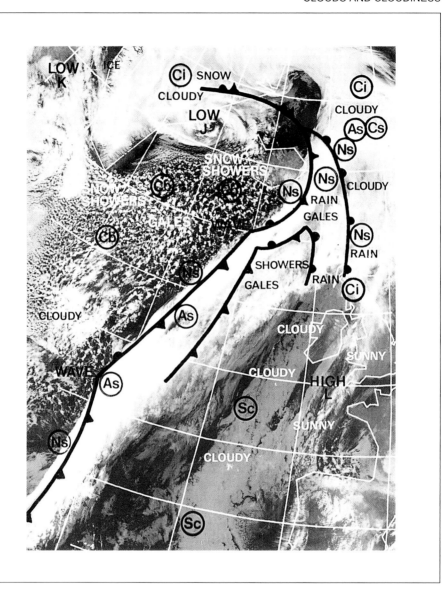

*Fig 3.2 What the surface frontal and cloud patterns look like on a TIROS image. Sometimes double frontal systems occur when other airmasses get drawn into the picture.*

we can see where one bit of weather is compared to the rest. However we are earthbound observers and we have to ask what we might see from the ground. Some help in this has been given by writing in the cloud types where they are to be seen – but what about the breaks?

Breaks are what we are always looking for. When it has been raining we will be anxiously scanning the windward

31

horizon for the signs of light that indicate a clearance.

Look at the region of the cold front in Fig 3.2 between the two waves where it indicates (As). Behind the edge of the frontal cloud there is a wide open slot where it will be sunny. It may not look much on the picture but it is practically as wide as Ireland. As the weather flies, that means perhaps three hours of sunshine before the showers break out. We have already seen that the field of speckles between Greenland and the front means many wintery showers and as the winds are gale force so it will be very miserable weather indeed up there.

We can detect a certain period of clear weather behind the clearing cloud of the front almost all the way along its length and this is what we most often find. For example it has been bad weather for some time but then the forecast indicates that a cold front will clear the air. As that front passes there will be a definite edge of cirrus clouds and then the sun will come out. Make the most of it because what normally follows is a build-up of heap clouds which often become big enough to produce showers.

Another aspect of holes in the clouds is seen where the cloud shield of Low J is 'eaten into' north of Iceland. What ought to be deep Ns clouds is found to be quite a substantial ration of sunshine. This is an example of the problems of the forecasters. By all the rules, that hole should not be there but it is. Some rogue has melted the clouds and that rogue is sinking air.

Depressions are regions where the air is mainly rising – and rising, wet air leads to deep clouds and rain or snow. Yet sometimes, and often inexplicably, some sinking air gets into the picture so that the clouds disappear and the sun comes out. For those in the sunshine it is very difficult to realise that only perhaps fifty to one hundred miles away it is raining heavily from leaden skies. However we see from the cloudscape that we cannot trust this sunshine to last for very long. One thing we must realise is that:

*When the weather is cyclonic you cannot trust clear spells to last very long.*

By cyclonic we mean weather in the circulation of depressions where more air is rising than sinking as opposed to anticyclonic where the air is mainly sinking from very high up.

Because rising air cools and condenses, its moisture turns into clouds and precipitation of all kinds, so cyclonic weather is going to always be tending towards the worst. What might be just a cloud bank under anticyclonic isobars becomes a period of rain or showers under cyclonic ones. Troughs produce heavy rain and showers when the isobars are cyclonically curved whereas they may just revert to drizzling when they lie under anticyclonic bends in the isobars.

### When the isobars bend

What do we mean by cyclonic and anticyclonic isobars? We can best answer this question with a weather map. (Fig. 3.3)

When you look at a weather map the rule is:

*Where the isobars curve in the same sense as round a low – the curvature is cyclonic. Where they curve in the same sense as round a high – the curvature is anticyclonic.* Look at the ridge of high pressure ahead of the low (L). Look at the first isobar out from the one that encloses the high. Around (a) the isobars are strongly curved in the same sense as round the high itself but at (b) the curvature goes the other way. At (a) we might expect clearing skies because the air is sinking strongly there. The curvature at (b) is cyclonic and there may well be more cloud there or even some light rain or snow. Where this isobar goes through the warm front it is weakly curved in the same sense as the other isobars that enclose the low. So as the curvature is weak, we expect cloudiness from the front but maybe nothing more.

Now looking at the isobars round the low we find them strongly cyclonically curved where the trough is at (c). There we must expect showers or longer outbreaks of rain whereas between (c) and the cold front the isobars are weakly anticyclonically curved and so there the weather, while still mainly cloudy, may well be much more benign.

*Fig 3.3 To illustrate the idea of cyclonic and anticyclonic curvature of the isobars and the wind changes to expect when the shipping forecast mentions 'cyclonic'.*

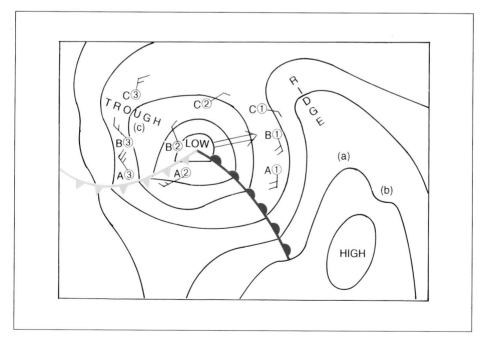

These examples will enable anyone to pick out which way the isobars are curving over their own area and so make sense out of maybe a period of better weather in a run of bad or conversely an even nastier bit than there has been previously.

## Cyclonic

Yachtsmen listening to the shipping forecasts will find the term *cyclonic* used in another sense to the one above. Because lows can be comparable in size with the forecast areas so, when their centres track across a sea area or a neighbouring one, the wind shifts that occur are far too complex to be described. Thus the mariner is left on his own to divine from established met theory what will happen. For example a yachtsman at A(1) sailing west has a southerly wind. He has to expect that when a warm front has passed the wind will veer (shift clockwise) to SW A(2). Of course as the low is tracking eastwards the situation A(2) will soon be transferred to A(1) with all the attendant changes of wind and weather that accompany the passage of a warm front. The yacht will have moved a bit as well but no matter which way it sails the changes of a travelling cyclonic situation will eventually overtake it.

Eventually A(2) has to be replaced by A(3), again with a wind veer – in this case to NW (which is typical). He will also have to contend with the, often nasty, weather associated with the passage of a cold front. These changes are the ones to expect when the forecast says 'cyclonic' and the yacht is to the south of the track of the low.

The yacht at B(1) is on the track of the low centre. His wind changes will be different. He will experience the 'eye of the storm' changes where the wind will fall lighter and lighter. Eventually it will reduce to nothing if he is fully in the centre of the low and then the wind will, B(2), begin to pick up from the NW or a similar direction. Eventually B(3) the wind will increase fully and showers may well break out. If he experiences the trough, which is swinging down round the low like the spoke of a wheel, those showers could well be heavy with squalls and even a rumble of thunder.

The yacht to the north of the centre C(1) will expect to have an easterly wind which, as the low centre passes by to the south C(2), will back round gradually to end up around north. Winds to the north of low centres are often much lighter than to the south. However when eventually in the relative position C(3) he must expect the showers as well. There are not always going to be showers behind a retreating low but there often are.

(Note; the previous refers to the Northern Hemisphere (NH). It will refer to the Southern Hemisphere (SH) if we make the changes given under the latter title in the Reference Section)

## Cloudiness

The amount of cloud that anyone experiences depends on many things. On coasts that face oceans such as a great deal of Atlantic Europe, there will on the whole be more chance of cloud than sunshine. As the moist ocean air moves inland it sheds much of its moisture and becomes drier.

Mountain or hill ranges eat clouds and rain so that the areas in their lee are much drier than elsewhere. It is the slopes facing the wind which get the worst of it, with sometimes days of low cloud, rain and drizzle, while the areas in their lee have at least broken sunshine.

Cloud satellite pictures on the TV will show where areas of persistent cloudiness lie. Even in relatively fair situations great islands of cloud tend to form and drift with the wind. These have a much greater chance of breaking up over land in summer because of the way the sun will 'burn-off' the cloud. In winter there is little effect of the sun and the clouds just remain and often intensify.

Obviously fronts and depressions bring much of our cloudiness but often these weather systems are on the move so that the breaks will come eventually. It is in slacker situations (often associated with the anticyclones) that the long periods of cloudiness, which we find so depressing, come along.

The sun on the earth leads to heap-cloud generation. A clear, often cool, early morning will soon become filled with Cu (and sometimes Cb) clouds produced by the sun warming the land. In fact in cool unstable airstreams the nights are often filled with stars as bright as jewels. Yet that gives the sun full rein in the morning to inject heat into the ground which will lead to thermals and so set off the clouds.

This process will be particularly potent when a cold front has passed last evening or maybe during the night. The amount of moisture left on the ground in the morning is enough to feed great masses of ragged heap clouds that form quite early. However very often it becomes less cloudy later. This follows because the sun has by now had a chance to dry up the ground so depriving the clouds of their fuel.

Another way in which an airstream becomes cloudy near the ground is by *turbulence*. Turbulence cloud forms when the wind increases and stirs up the first two thousand feet or so. The eddies formed by the wind lift moist air and so ragged layer clouds form especially over moderately hilly country. There is a limit to this process however, if as happens occasionally, the wind blows at above gale force over land, the depth of air mixed by the turbulence includes dryer air higher up and so the howling wind blows under clear skies.

A considerable amount of knowledge of what to expect on any day comes from having an idea of what kind of airmass you are in and what usually happens to that kind of air either over the land or over the sea depending on where the wind is blowing from. We will give some ideas about this in the next chapter.

Chapter Four

# From the Four Winds

When most of us live in the temperate zone where the weather is always on the change, it is sometimes difficult to realise that there are regions not too far away where the weather hardly ever changes.

These areas are the centres of permanent anticyclones and they are the places from where the airmasses that make the weather originate.

An *airmass* is a vast block of air with the same general characteristics and it got these characteristics by sitting for weeks or even months over one locality. By doing that it acquires warmth or cold, dryness or humidity through its entire depth. Then, because no weather system is impervious to some changes it gets drawn out of its *source region* and transported maybe thousands of miles to clash with another airmass of different characteristics. From the encounters of airmasses come most of our depressions and the fronts that are an integral part of them.

### The world's weather systems

Figure 4.1 is a plan of the world's weather systems reduced to their absolute basics. It is a plan on which to hang your hat but not one to rely on for details. The Southern Hemisphere (SH) is largely a reflection of the Northern (NH) so we can describe the latter and in so doing include the former. Through the

low pressure zone of the temperate latitudes wriggles the *polar front*. This is the more-or-less continuous line of division between cold air from the poles and warm air from the tropics. It lies in a weather zone of constant change.

To the south lie the permanent anticyclones of the Subtropical High Pressure zone and to their south, over the Equatorial region, there are the Doldrums so feared by the sailors of old. There is a kind of front here also. It is called the *Intertropical Convergence Zone*. It is not a front like the polar front but it divides the air of the SH from that of the NH.

To the north of the Temperature Zone lies the Polar *High* – a more-or-less permanent region of very cold and relatively humid air. It is an important member of what are called *source regions*.

### Source regions

What your own particular source regions are depends on where you live but Fig. 4.2 enables you to make up your mind for any locality. As the airmasses that are separated by fronts have to be different – one cold and one warm – so what we are looking for are polar sources and sub-tropical sources.

Take Atlantic Europe. There are two predominant source regions which bring

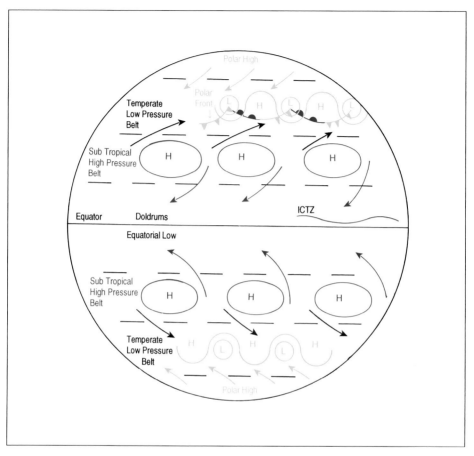

Temperate
Low Pressure
Belt

Sub Tropical
High Pressure
Belt

Polar High

Polar
Front

H

H

H

H

H

Equator      Doldrums                    ICTZ

Equatorial Low

Sub Tropical
High Pressure
Belt

H          H          H

Temperate
Low Pressure
Belt

Polar High

wet air and both have to come from sources over the sea. So the cold, wet airmass is from sub-polar seas and is called *maritime Polar* (mP). The warm, wet airmass comes from the Sargasso Sea area of mid-Atlantic and is called *maritime Tropical* (mT).

We see maritime Polar air at its purest on the NW wind coming down behind retreating cold fronts. Then the air is clean and visibility for the most part is excellent. Over the sea showers may occur by day and night while over the land the nights are usually clear and the days build into showers.

*Fig 4.1 The weather systems of the world. There are three low pressure belts; a temperate one for each hemisphere where all the most changeable weather occurs plus an equatorial one where weather is in some ways more unpredictable. There are highs over both poles as well as over the sub-tropics of both hemispheres and the latter supply the Trade Winds and the temperate low pressure belts with warm moist air.*

We see maritime Tropical on the SW wind. Then clouds are low and

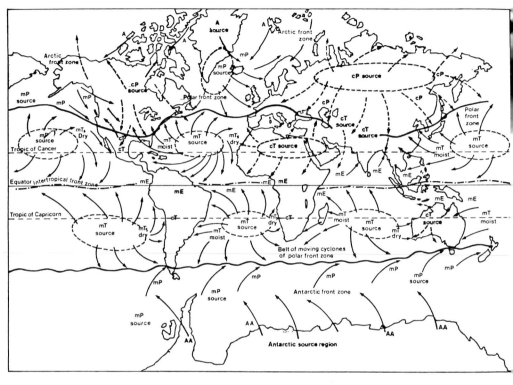

*Fig 4.2*
*The sources and names of the airmasses of the world. In this diagram the important returning maritime Polar (rmP) airmass which brings good weather from the West to Atlantic Europe is not shown.*
*Key*
*mT=maritime Tropical; mP=maritime Polar; a suffix s means the airmass is subsiding; E=Equatorial; cP=continental Polar; cT=continental Tropical; A=Arctic; AA=Antarctic.*

abundant. It drizzles and rains as the very moist, warm air is lifted up over coasts and hills. Visibility is poor and may even give rise to fog especially on coasts and over hills. It could not be more unlike mP air.

It is this very contrast between the two airmasses that leads to the frontal weather where rain is the order of the day. The only element that the two airmasses have in common is humidity. They are both crammed with moisture just waiting for the correct conditions to deposit it on us.

Dry air comes to Atlantic Europe from one cold source – the Siberian High – and one warm source – the North African High. Following the system of lettering airmasses (as with mP and mT air) the airmasses from within dry continents are going to be *continental Polar* (cP) and *continental Tropical* (cT).

Continental Polar air comes to Europe largely on the E or NE wind and it can be benign or, in winter, the cause of much snowfall when a ridge of high pressure from the Siberian High juts out across Scandinavia. Low pressure systems to the South feed warmer air into this persistent and extremely cold blast from Russia, leading to a very bad winter. At other times of year the dry easterly can be very pleasant.

Continental Tropical air comes mainly on SE winds and it often provides unnatural warmth in the spring and heat waves in the summer. It is the airstream which deposits Saharan dust on our cars in desultory showers and it is relatively rare just as is cP air.

Thus here we have the four winds for Atlantic Europe:

NW = wet and cool
SW = wet and warm
SE = dry and warm
NE = dry and cool.

In practice of course there are many variations on this basic theme. How often, for example, do we get cool SE winds blowing in and gathering strength before the onslaught of a major depression? The NW wind may become a cool or cold northerly or even north-easterly. The SW wind, which normally does not produce showers, may bring in quite hefty showers when it is due to a low not far away to the west. Or it may bring thundery showers in the spring, summer and autumn. The NE wind can be east-erly or even south-easterly and it all depends on where the big anticyclones have decided to rest.

**Blocking highs**

Anticyclones (or highs) in general do one of two things – they either travel or they block. Travelling highs fit into the merry-go-round of lows that tend to circulate around the temperate lattitudes of either hemisphere and move along with them. They bring a day or two of respite from the misery of rain, wind and cloudy skies when a succession of lows cross an area.

Blocking highs on the other hand do what their name implies. They sit four-square over one locality and block the way for the lows which are forced to go elsewhere. They may bring weeks of one type of weather and it is often weather uncharacteristic of your locality.

If the highs are not in their accustomed localities, then at the same time the lows too have to go to unusual places. Lows cannot penetrate blocking highs and must visit the areas which lie on the high's flanks. Thus, as suggested in Fig 4.3, a high over the seas south of Iceland, which is the traditional place for low pressure, must push unseasonable weather into Spain and the Mediterranean.

In Fig 4.3 another high situation is shown off the coast of Newfoundland. This high may well be an extension of the permanent source region in mid-Atlantic replacing the many lows that thread through the area. Now also, wet and warm air is brought up from more

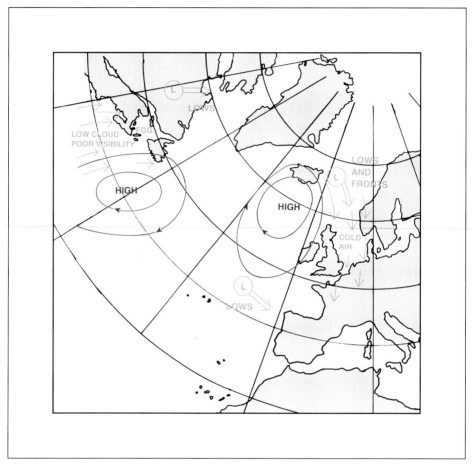

*Fig 4.3 Highs centred for lows and fronts to bring bad weather from the North into Europe and for mT air to bring poor weather into northern America.*

southerly seas and so the American East Coast is plagued by low cloud and fog. The lows will now be tracking north across northern Canada and into Greenland while others will break away to travel down the central and western U.S.

The normal summer situation for Atlantic Europe is seen in Fig. 4.4. Now the weather comes on normally light or moderate westerly winds and it is good for the beach. The weather is not too hot – that is usually reserved for winds with something easterly about them. It is a time when seabreezes abound round the coasts taking over and modifying the daytime winds.

Now the lows are going on their accustomed paths through the seas south of Iceland and so bringing poor weather to Scotland and Scandinavia generally. There is a block but the weather situation is normal.

Looking across the Atlantic, where in Fig. 4.3 we had lows, we now have a high. In this case the eastern US experiences cool or even cold, wet easterlies. Any fronts or lows that get into the circulation of the high can bring some nasty unseasonable weather.

*Fig. 4.4 A ridge from the Azores anticyclone brings fair weather into Europe while one centred between Labrador and Greenland brings cool easterlies into northern America.*

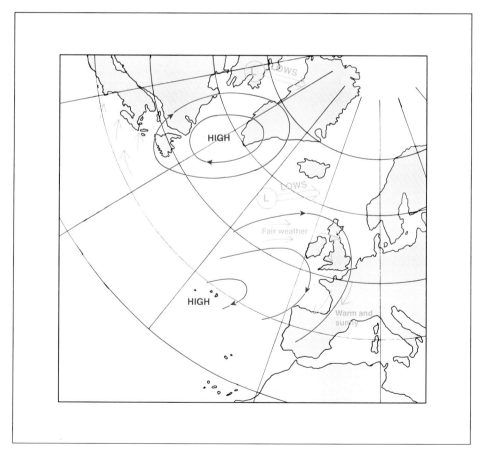

## Sources for America

The airmasses that contribute to the weather of northern America have the same names as in Europe. Fig 4.2. The polar front is fed with warm moist mT air from the same mid-Atlantic source region that feeds Europe but now it is south-easterly rather than a south-westerly. This air is the most prevalent of all on the east coast in summer and leads to oppressive heat and humidity as well as massive outbreaks of thunderstorms. On the west coast the cold, wet mP air meets mT that is not so wet as its European counterpart.

On the east coast the mP air comes much dried out by passage across the continent but there are still plenty of lows formed along the polar front as well as over Alberta. The great land area of Canada forms a continental polar source which gives dry cool air in summer or dry, extremely cold, air in winter.

## Other airmasses

It is important to mention two important kinds of airmass that are not included in the basic four above. Of these the most prevalent is the polar air which comes to Atlantic Europe on a very long track from way out in the Atlantic. It is forced onto this long trek by semi-permanent low pressure south of Iceland. It is called Returning maritime Polar air (RmP) and it brings some of our most benign weather. The wind is usually W and the clouds are humble cumulus or maybe strato-cumulus. The air has lost much of its moisture and has developed an *inversion* (which puts a lid on the cloud tops) so showers are very rare. In summer it brings us ideal holiday weather and is likely to be pleasant at any other times of year also.

Arctic air is quite the opposite. When the isobars come straight from the arctic then the weather is very cold, often very wet and thoroughly miserable. A special kind of frontless whirl of air called a *Polar Low* develops in this nasty airstream and brings wintry showers or longer periods of cold rain. Polar lows are not at all pleasant.

You might be forgiven for thinking that an airmass called 'polar' ought to be colder than one called 'arctic' however the difference lies in the way they are modified by the sea over which they travel. Polar air has usually had time to acquire warmth from the seas as it comes South on a longer track than arctic air which comes direct from the North.

In this run-down on the world's weather systems and the airmasses that affect the temperate regions we have had to give just the basics. However basics are what we need to hoist-in when we first try to become acquainted with the vagaries of the weather. In the next chapter we are going to explore the way we can do some forecasting for ourselves when the weather is changeable.

Chapter Five

# The Message in the Winds

While we are usually told to look to windward for the weather, it is a fact that most often the upper winds on which the bulk of the clouds come are not in the same direction as those at the surface. This is particularly so when the weather is on the change. Then the winds in which the cirrus clouds ride are often moving at right angles to the wind in which the cumulus clouds ride.

Much of the output of the computers of the world's weather services is dedicated to producing actual and fore-cast charts of how the upper winds are moving or are expected to move. These are called *contour charts* because they show meteorologists how high up a particular pressure level can be found.

While surface pressure is on average around 1000 millibars (mb), at the heights at which cirrus clouds form the pressure has fallen to about a third of this. The contour charts are drawn for standard levels of which the 500mb and 300mb are of special importance to us.

The 500mb contours show the way the pressure systems move and the 300mb contours show the way the cirrus clouds move. Looking carefully at cirrus is the only way the layman can observe for himself what the high-altitude winds are doing. Thus there is a great deal to be gained from looking at the shape, speed and direction of movement of the 'mare's tails' when it comes to doing your own forecasting.

So it is to the 300mb contour chart together with the normal surface chart that we must turn if we are to understand a simple method of forecasting the coming weather. I call this the '*Crossed Winds*' method because only when the winds at cirrus height are crossed in direction to those at cumulus height can the weather be really on the change for better or for worse. However before we can explain all this we must put down some basics of the weather map.

## A model depression

Some of the way depressions form on the polar front and then live out their life cycle from birth, through their virulent youthful stages to eventually drift into old age and die out is evident from Fig 7.4. Just for now we can draw a very simplified version of a maturing low as seen on the weather charts. (Fig 5.1)

This diagram contains the important facts about a low. Firstly the winds will be seen to blow round the low centre keeping

*Low pressure on their Left (Note the mnemonic of L in Low and Left).*

The low centre moves from West to East or from SW to NE. Not all lows do this but most of them do, so this low is typical.

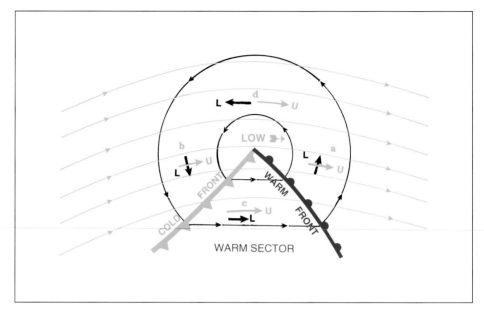

*Fig 5.1 A model depression with the winds at cirrus height to illustrate the Crossed Winds rules.*

The low has two fronts enclosing a vee-shaped warm sector. The low therefore has two major airmasses. To the north of the fronts the air is of polar origin while in the warm sector it is trop-ical origin. The warm front is the surface of separation between cool air ahead and warm air behind, while the cold front has cold air behind and warm air in front. Fronts are therefore named as follows

> The air behind the front gives its
> · name to the front.

As a general rule the isobars in the warm sector tend to be straight and the low moves parallel to them.

For our present understanding this is all we need, so we will turn our attention to the winds at an altitude of some 4 to 6 miles (6–10km). The fact that you cannot be at all sure how high this is from the ground need not be a stumbling block. If you recognise cirrus clouds you must be looking at the height we are describing.

## Upper winds

These tend to blow more strongly than surface winds but not everywhere or always. However, for the situations where bad weather is involved we can say that the winds at altitude will be blowing very strongly indeed. Their greatest strength is reserved for a high-speed 'river' of wind called a *jetstream*. Jetstreams are the friends and foes of trans-Atlantic jet aircraft. When the

100–150 knot jetstream winds are behind them then they cross from America to Europe in record time but on the way back they have to operate at a different altitude to a jetstream so as to steer clear of its worst effects.

The general direction in which upper winds blow is from west to east and you will see that this keeps the pole on their left and the tropics on their right. Thus in general we have another rule for the high altitude winds –

*Upper winds blow to keep Low temperature on their Left.*

However Figure 5.2 shows us that the upper winds are always snaking about and even going round in circles, so they are often not blowing from west to east. This does not stop the upper westerlies taken as a whole from doing as their name implies – blowing round the Earth from a largely westerly direction.

**Contour charts**

We can best explain about the winds at cirrus levels if we look at Fig 5.2 which is for winds around cirrus height at midnight 4 November 1985. In order not to confuse the diagram no land masses

*Fig 5.2 The contours at cirrus height (300mb).*

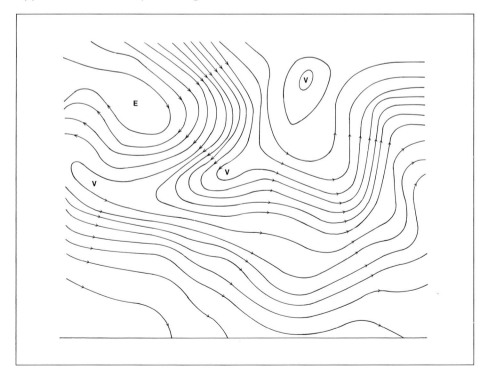

have been included but they are the same Atlantic ones as in Fig 6.1. Not putting in the land also emphasises that this chart could have turned up anywhere around the Northern Hemisphere. The surface weather can be highly modified by the land or sea over which it travels but the winds at high altitude are not greatly affected by the terrain below them.

To find the jetstreams look for where the contours crowd most closely together. A major one comes swirling down from the top (which is northern Canada) to make a sharp bend round a 'valley' in the contours (V) (which is over mid-Atlantic) before storming on to the north of Scotland and so across Scandinavia to northern Russia. This is the *Temperate Latitude Jet.*

Further south there is another jet – much less contorted than its colleague further north and not so strong. This is the *Sub-Tropical Jet.*

Because these charts are 'contours' so they can be envisaged as representing how high or low is the surface where the pressure is 300 millibars. We have to imagine the surface like one of those three-dimensional maps of the terrain which sometimes appear in geography classes. The contour map shows (E) where the pressure surface domes up indicating mainly warm air below it and also where it dips down (V) into troughs and valleys. In the latter cases there is mainly cold air below the points in question.

When we visualise the 300mb 'surface' this way we see how true it is that upper winds blow to keep low temperature on their left and so high temperature on their right.

### Winds over lows

As we saw in Fig 5.1 the warm air in a low is to the south of the centre and enclosed between the fronts. Thus to obey the previous rule the upper winds have to blow in an arc around the northern part of the low's circulation to try and enclose the warm air.

This way we can add contours of upper wind to our simple depression model (Fig 5.1) and find that only in some places are the winds below and above in the same direction. For most of the area under the influence of a low the winds at altitude blow across those being felt on the Earth below. Thus we can formulate some rules to tell us when a warmer or a colder airmass is on its way.

### When warmer air is coming

If we can make reliable weather observations that tell us a warm airmass is coming then the second message is that there must be a warm front separating the two airmasses. That in its turn means gathering cloud, rain or snow, low stratus on the hills and almost certainly increasing wind. Thus if we formulate a *Crossed Winds* rule for coming warm air it tells us vastly more than that single fact. It gives us intelligence about a great mass of weather over a very wide area. It may also forewarn us many hours in advance of the dire things that may follow. That must make it well worth studying.

We can see at 'a' in Fig 5.1 that if we label the upper wind U and the lower wind L, then L and U are at right angles (or nearly so) but they have a certain orientation which is summed up like this:

*Stand back to the lower wind and the upper wind will come from your left hand when a warmer airmass is coming.*

But how do you know the upper wind direction?

Answer; through observing the cirrus clouds, which stream in ahead of warm and occluded fronts.

**Watching the cirrus**

Cirrus clouds are unique in the sky. No other cloud looks like cirrus and this is because of the way they are formed. In Photo 5.1 we are looking at some out-

*Photo 5.1 While not entirely typical, these fall-streaks illustrate the way that many such clouds can combine their streaks together to make cirrus banners.*

*Photo 5.2  Spectacular jet cirrus.*

riding cirrus fallstreaks streaming back into a mass of cirrus in the background. Below we have cumulus cloud which has begun to die out as the upper cloud moves in. This is a typical skyscape when a major front is in the offing.

We have to recognise a certain feature of skies that have a prognostic value and that is organisation. Not all skies will have a definite recognisable direction (or set of directions) but those that do, have in my experience, something to say about the coming weather.

Here in Photo 5.1 we see how the fallstreaks lead our eye in the same direction as the lines we can detect in the cirrus mass behind. What we are looking at is almost certainly *jetstream cirrus* and it has the linear appearance which we are looking for when we wish to detect this particular form of cirrus which foretells the worst weather.

Another important aspect of jet cirrus is the way it has a definite edge on its polar side i.e. in the Northern Hemisphere, its northern side but its southern side in the Southern. It is formed by the amalgamation of the fallstreaks of thousands if not millions of individual cirrus clouds. These combine into great banners which lie more-or-less parallel to the winds at their level and a spectacular example is Photo 5.2.

This is the classic 'windy sky' which has spawned all the lore about mares tails and in more modern times a song that speaks of 'rows and flows of angel hair'. This sky obeys the principle of linearity to a remarkable degree and in

effect says 'Look at me and do not ignore what I say'. Of course the convergence towards the horizon is just the same effect of perspective as we get when we stand in the tracks and look down a railway line.

So here we are ignoring the fussy little details along the edges of the banners and concentrating on the linear form of the whole upper sky. And it is of course so marvellously contrasted with the blue sky because it has that pure whiteness of ice clouds. If this cloud should be thought to be a lower cloud composed of water droplets then it cannot be because the latter would show shadows. There are no signs of shadows. Just occasionally, when the coming weather is likely to be of the most violent kind, the density of these jetstream cirrus banners becomes such that they can cut out the sunlight and so show some shading of their bases but such banners are rare, thank goodness.

By now we have enough clues to be able to recognise jet cirrus when we see it but there is another aspect which can help to make it more certain that the winds at cirrus height are blowing at 100 plus knots.

Because they are so high it is often difficult to detect the motion of cirrus clouds without measuring their movement against some fixed object like a TV aerial or a lamp-post. If you have to do that then you are not dealing with a strong jetstream and the coming weather, while it may well deteriorate the situation, should not blow up to full gales.

However if you concentrate on a cirrus element or two and can unaided see them moving against the background of the sky you are certainly looking at clouds carried in winds of a hundred knots or more. That is when you begin searching the airwaves for a forecast. The link between high-speed upper winds and strong surface winds later lies in the fact that the stronger the jet the more virile are the depressions that form in association with it.

One final clue is the way the cirrus elements form hooks. Hooked cirrus comes about because the wind in the jet core has to unwind into the surface wind in a mere few miles of depth. (See Fig 7.2). The cirrus heads form just on the equatorial side of the jet core and their fallstreaks must fall into rapidly decreasing wind strength. Thus the tails get left behind in the slower wind while the head flies on. If the fallstreaks are therefore sharply hooked back they indicate an extremely strong wind regime up there and help confirm that what you are looking at is jet cirrus. Conversely fallstreaks that sink almost vertically down the sky indicate little wind speed aloft. Such fallstreaks often accompany skies ahead of thundery out-breaks.

There are, of course, other cirrus skies where the individual cloud elements are not numerous enough to produce the impressive banners of jet cirrus. They indicate that the air at altitude is less moist than when banners form. That in its turn indicates that the depression as a whole is probably less active than one where the mares tails sweep across the sky.

In this introduction to the *Crossed Winds* method of forecasting many of the details are being glossed over and we shall return to them in later chapters. For now, so as to tie up the method let us see what it tells us about.

### Better weather coming

The weather has been foul for a time due to a depression tracking by to the north but eventually the clouds clear into blue sky. As the cold front cloud clears it often does so with a clean-cut edge along which cirrus clouds can be seen to be moving. This cirrus is, we observe, coming from around SW and we remember that before all this nasty weather blew up the cirrus had been winging down on NW winds. Because of the way the upper winds have to curve round the warm air we see why this is (Fig 5.1). At 'b' behind the cold front the lower wind has shifted to NW while the cirrus, riding the upper wind U, streams along the back edge of the cold front's cloud from SW.

We now see that the winds are again crossed more-or-less at right angles but in the opposite sense to those ahead of the depression. This gives us another *Crossed Winds* rule but this time it tells us that a colder airmass is coming. The rule now is –

*Stand back to the lower wind and if the upper wind comes from your right hand then colder air is coming.*

You might say that cold air is what you

would expect behind a retreating depression but what if another depression is lining up to the west to take over where this one left off?

In that case there would be a brief respite that might tempt the yachtsman out of port. The landsman might plan a family outing for tomorrow thinking that the worst is over, only to be disappointed when the next morning he wakes up to overcast skies and a gathering wind.

The *Crossed Winds* rule above tells us that as a cold airmass is coming in then we ought to get a day or two at least of more settled weather. Sometimes this orientation of upper and lower winds precedes the build-up of a good anticyclone to the west which will break the run of depressions and give a week or more of better weather.

However if the two winds are not crossed in a positive way as the cold front recedes we do not believe the clearance and seek information on what else may be in store.

Finally note that at 'c' and 'd' the winds are more-or-less parallel. At 'c', in the warm air, there may be cirrus to be seen through gaps in the cloudiness which pervades warm sectors and it will often be moving along with the surface wind direction or nearly so.

To the north of the low centre, at 'd', the two winds are blowing in opposite directions. It will depend on where you are compared to the edge of the cloud shield that lies over a low centre whether or not you can see any upper cloud. Assuming you can, then an observation of cirrus moving in a mainly westerly direction, while the wind you have blows from the east, will say at once 'There is a low tracking by to the south'.

In order not to overload this chapter much useful detail has been left to the next and subsequent chapters. However, enough has been said to enable the budding forecaster to get out in a period of changeable weather and start looking at the high clouds. Observe the way they move and compare that with the surface wind. Then apply the rules and see what subsequently happens. However, for this to be a reliable method we have to know somewhat more about pressure systems and their winds.

Chapter Six

# Weather Systems

Weather systems are great masses of weather organised into some kind of recognisable and understandable arrangement. So lows and highs are weather systems but under this heading we are going to include fronts which are important parts of lows. Fronts also swim

*Fig 6.1 The surface isobars and fronts for the same time as shown in Fig 5.2.*

into the circulation of highs but by then they are usually well past their prime.

The model low introduced in the last chapter was sufficient for our purpose but the variations on such a simple theme are legion. We can best illustrate this by looking at an actual weather map. This one (Fig 6.1) is the surface chart which goes with the contour chart (Fig 5.2) in the previous chapter. This time we

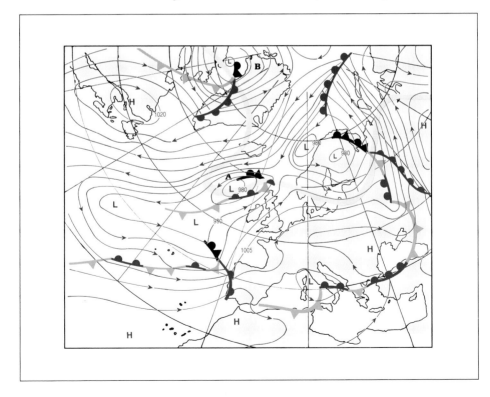

need the land surfaces so that we know where all the details are situated.

Strangely enough on this particular chart there is not one low which truly resembles our model low but they have recognisable features. This is also an object lesson on real weather systems. They may have most of the features outlined in the last chapter but in reality very few look like the text book.

**Occluding depressions**

The first reason why the lows off Scotland (A), and north of Greenland (B) do not look like the model is that they are occluding. This means that the cold front of the young depression has, as time has gone on, overtaken the warm front and in so doing forced the warm air off the ground altogether. Occlusion is a natural part of the life-cycle of depressions but once they begin to occlude they are past their prime and have usually lost much of their earlier fire.

However we cannot write them off because sometimes new lows form on the point of occlusion (i.e. where the warm and cold fronts meet) and can get locally nasty.

Low A has driven its fronts a long way on. Strictly by the text-book they ought to look like those of Low B but modern charts are drawn using the intelligence from satellite cloud pictures etc and the forecasters work on the premise that 'if it's there, put it in'. In pre-satellite days it was often impossible to know what a weather system out over the ocean wastes actually looked like and so a text-book representation was put in roughly where it was thought to be.

The warm front of Low A can be seen to have largely died out while the cold front sweeps back not far south of the low centre. It has developed a wave and in the area of the wave-tip weather will have deteriorated markedly. The wave cannot develop into another depression because it is too close to its parent but it can add to the sum-total of weather associated with Low A.

Low B is much better behaved. It looks just like the diagrams of occluding depressions in the text-books. It still has a wide warm sector and the winds shift direction clockwise (veer) across the fronts as they should do. On this map projection we have to rotate the chart and look at it as if it were on a Mercator projection. This means that to see the proper directions for Low B we must rotate the chart until Greenland looks right. Then we see that the winds ahead of Low B are SW. Those in the warm sector lie between SW and S while those behind are between W and NW.

The occluded part of the front starts in the centre of Low B and is aligned with the cold front. That makes it a cold occlusion. If the occlusion and the warm front had formed one continuous line then it would have been a warm occlusion. The occlusion of Low A is a warm occlusion. The distinctions between warm and cold occlusions are largely academic for us. Occluded fronts may have, ahead of them, many of the features of warm fronts and behind them those of cold fronts but both are usually less active than the pure fronts from which they formed. (See occlusions in the reference section)

An even more odd frontal system is the one of Biscay. In this case there is no apparent low centre and the occlusion lies in a trough in the isobars. The wind shift as the occlusion passes will be quite marked from between SW and S to around NW. The trailing cold front has not just one but two waves on it, both of which will have locally deteriorated the weather.

To look at the frontal system over Russia we have, again, to rotate the chart so we are looking North. We have over northern Scandinavia a 'two fried egg' low. Such twin systems do not move very far and sometimes rotate round one another. They are regions where the weather systems have stopped moving on round the hemisphere. We should now recognise this occlusion as a cold one as it follows the line of the cold front. However once systems move well into a continent they tend to become complex and difficult to place in any well-defined categories. We would be advised to ignore such systems for the present and just concentrate on more obviously text-book arrangements.

**Warm fronts**

If anyone asks me what is the first thing they should learn about met I can never think of anything more important than the cloud sequence which runs ahead of warm fronts and occlusions. Recognising

the succession of cloud types that precede the fronts also foretells that a depression is in the offing with all the weather that entails.

Frontal surfaces always slope. Cold frontal surfaces generally slope up twice as steeply as warm frontal surfaces. This means that on the whole cold fronts clear twice as quickly as warm fronts.

A cross-section of an active warm front in three dimensions (Fig 6.2) shows where the cloud types can be seen. This diagram also gives typical times and distances for well-developed warm fronts that are part of active depressions. These active fronts where the air is mainly ascending are called *ana fronts* as opposed to old dying fronts which are called *kata fronts*. In the latter case much of the air over the front is descending leading to dispersal of upper cloud and the cessation of precipitation. We are, for now, only going to discuss ana fronts.

**Signs in the sky**

Often the first signs of a warm front need to be noticed through the gaps in the cumulus that has formed in the fair weather situation that usually precedes bad weather. In Photo 6.1 the cirrus is fairly obvious but it is not always so. Even before the cirrus shows up, unusually dense trails from aircraft can indicate that the high regions are primed with moisture and so a front is probably beginning to move in.

There is such a trail cutting down across the lie of the cirrus in Photo 6.1 and trails are very useful when it comes

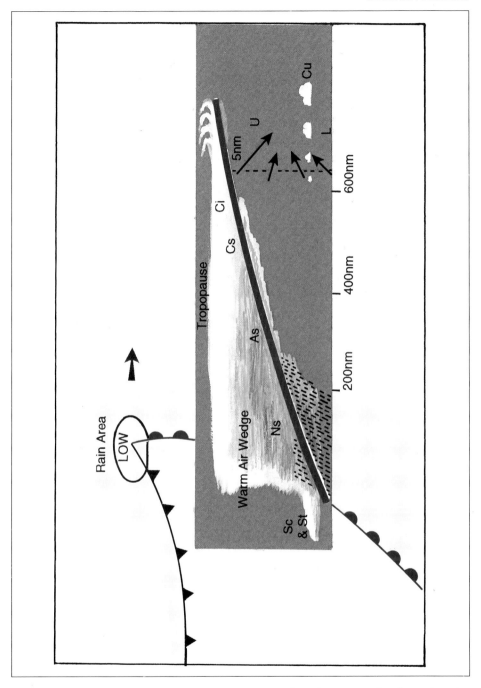

*Fig 6.2 3-D cross-section through an ana warm front. The way the winds unwind is shown on the far right. The distances are typical.*

*Photo 6.1 Very often you can use the flight of cumulus cloud as a true direction for the lower wind and compare it with the upper wind as gauged from the cirrus. Aircraft trails as in the middle of the picture help with the latter.*

to seeing which way the wind is blowing up there. The direction the Cu moves is also the best indication of the lower wind when using the *Crossed Winds* rules outlined in the previous chapter. We have only brushed the surface of this method and have a lot more useful things to say about it in the coming pages. For now we will concentrate on recognising

the warm front from its cloud types.

The cloud that follows cirrus is cirrostratus. We cannot mistake this change because the sun or the moon will become ringed by the characteristic halo as in Photo 2.7.

When the grey 'inverted mud-flats' of altostratus move in, rain is much nearer but it may still be some hours before it starts. The trademark of frontal altostratus is the way the sun disappears more-or-less slowly into the murk as if it is going behind increasingly thick ground glass. Photo 6.2 shows a typical As sky ahead of an ana warm front.

As the As thickens (the sun had long

⇑ *Photo 6.2 The way we recognise altostratus when the sun gradually fades into it as it thickens.*

⇓ *Photo 6.3 The stratus cloud that is called pannus and forms in the wet air under nimbostratus that is just about to rain on you.*

since disappeared) and before rain is felt on the ground fine rain will be falling from the deepening clouds. The small droplets evaporate on the way down and so do not survive to the ground. This evaporation makes the layer below the cloudbase very moist. In this moist air a new cloud form occurs. This is technically called *stratus pannus* or just pannus and it has a useful role to play.

Pannus is seen forming under As in Photo 6.3 and its message is 'Reach for the waterproofs' because if thin rain is already falling then in only a few minutes it could be thickening and reaching the ground.

With well-behaved ana warm fronts the rain will now steadily increase in intensity and the pannus will spread to cover the entire sky making the visible cloudbase very low and increasing the gloom. The wind almost always increases with time and particularly when the rain really begins to fall.

If the rain comes and the expected wind does not materialise do not be lulled into a false sense of security. It is a well-known fact that when rain appears ahead of expected wind then the wind, when it comes, will be particularly strong. This follows because the rain clouds have been pushed on by high-speed upper winds which indicates a virile depression is in the offing and that will almost certainly spawn strong wind later.

Rain only falls continuously from nimbostratus clouds. There is no recognisable distinction between alto-stratus and nimbostratus from our point of view except that the As has thickened to such an extent as to be deep enough to produce rain. The answer to the question 'How do I recognise nimbostratus from altostratus?' is a simple one. When altostratus begins to rain, call it nimbostratus.

Thus we can now summarise the cloud sequence ahead of warm fronts and occlusions.

1. *cirrus*          *hooks, sheaves and mares' tails*
2. *cirrostratus*   *ring haloes*
3. *altostratus*    *sun disappears gradually*
4. *nimbostratus* *rain or snow (not drizzle)*

**The passing front**
When you have watched the cirrus give way to cirrostratus and that, in its turn, is lost behind gathering layers of altostratus you can be pretty certain that rain (or snow) is on its way. But when?

That is a difficult one. Without a doubt one of the chief problems for the professional forecasters is timing the weather features they are forecasting. They spend a lot of time and effort in trying to improve their performance in this direction but still the fronts will not always play ball. It may be that improved techniques will iron out this problem in the future but for now the person on the spot has to believe that the weather will come but not necessarily when the forecast says.

Research into the efficiency of timing of fronts done by the Meteorological

Office themselves indicates that fronts tend to turn up later than forecast. That is an error in the right direction but can the man-in-the-street do any better? The over-all answer is probably not but let us put down a few pointers to timing.

I have always used a rule-of-thumb that the time between the arrival of cirrus and the loss of the sun into altostratus is about the same as the time that will then elapse to when it rains. So if, as in Fig 6.2, we should first see the Ci around breakfast-time (i.e. 08.00–09.00) and the sun is disappearing into As at lunchtime (13.00–14.00) then it ought to be raining by early evening. The rain could well last about the same time i.e. about 5–6 hours in this case.

There is however an old saying about weather systems.

*Long foretold – long hold*
*Short forecast – soon past*

– which means that when the cloud build-up is rapid then the rain, if and when it comes, will also pass quickly. We are now talking about older less virile fronts and occlusions and younger more active ana fronts tend to take longer to pass.

### The warm sector

We see from Fig 6.2 that, while you may not be able to see it from the ground, when the warm front passes the deep nimbostratus clouds pass as well. After the edge of the front near the ground has passed, often with dark wisps of trailing stratus called *virga*, there may be a brief opening in the overcast but it often shuts

in again as soon as it appears. Now we are in the warm sector and the cloud will be low stratus and stratocumulus from which drizzle may fall. Whether that happens or not can rest on the terrain upwind of where you are. Drizzle is most likely over coasts facing the wind and anywhere where the ground forces the air to ascend.

In the warm moist tropical air it may rain on the windward slopes for hours and in some cases, where the cold front does not appear, for days. When the cold air is held back the reason is that high pressure has developed out to the west so forcing the cold front to remain to the north.

The feel of the air in a warm sector is 'muggy' but a great deal of this mugginess is due to moisture evaporated from the wet ground after the rain. Given time, and away from the windward coasts, the warm sector weather may dry out to pleasant warmth especially in the lee of high ground. In practice the weather in the mT airmass may be many things but it always tends to be more cloudy than not, warmer than you might expect and with high relative humidity. This means that the risk of fog at sea and over high ground is always present. It may also tend to fogginess over low ground in the winter half of the year. At some time or other the tropical airmass has to be replaced by a polar one. How long the warm air will remain depends on many factors not the least of which is where you are, compared to the point where the warm and cold fronts meet. If the

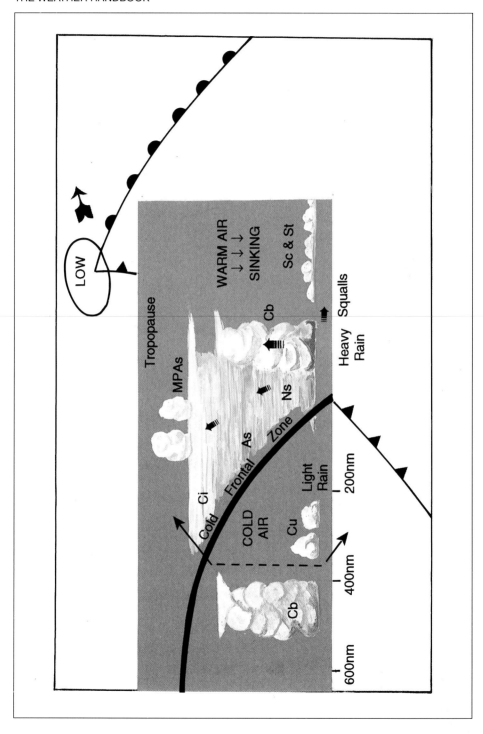

⇐ *Fig 6.3 3-D cross-section through an ana cold front. The distances are from where the cold front is first encountered on the ground. On the left the crossing of upper and lower winds for colder air coming is illustrated. The MPA's are mesoscale precipitation areas from which heavier rain falls into generally light rain.*

depression is occluding and you are under the occlusion then there will be no warm sector anyway. Warm front-type weather will give way to cold front-type weather without a significant break.

The further south you are the wider the warm sector but maybe the fronts will not be so virile. So little guidance can be given as to how long the warm air will be with you. A glance at a TV chart can help a great deal here.

### The cold front

The ana cold front can be a ferocious beast with sudden squalls and torrential rain. Peals of thunder may shatter even the winter night as a sharp cold front goes through. The reason for this ferocity is the way the cold air sweeping in behind the warm is lifting the latter off the ground at such a rate that great cumulonimbus clouds grow along the line of the front embedded in the nimbostratus of the front itself. (Fig 6.3).

This zone of strong weather passes relatively quickly to give way to much lighter rain which tails away to altostratus and eventually cirrus. It is here, on the trailing edge of the cold frontal cloud, that you can often make a prediction of

the weather to follow using the *Crossed Winds* rule. (p 50)

The very sharp updraughts along the front's leading edge must be compensated for by downdraughts ahead of the front proper so that you can get a short break in the cloudiness. This may enable you to see the towering anvil-headed cloud-mass bearing down on you. However it must be said that sometimes this does not happen and the first you know of the cold front is when it rains, hails or snows on you at the same time as it buffets you with squalls of wind. (See photo facing title page).

There will however be a veer of wind when cold fronts pass and when the low cloud elements appear to be going in two different directions at once, the different wind directions can sometimes be seen. This is an immediate sign of wind shift and also heralds a new seaway direction (if it has not already arrived). Coupled to the squalls from the Cb clouds this can be a potentially dangerous time for yachts at sea.

In this chapter we have described the weather of virile ana fronts. Most fronts that make their way into land masses lose much of their strength but they still retain memories of their former lives. They may even get injections of cold air which will give them new leases of life and so no front that appears on a chart can be quite written off. We will talk about such fronts and other systems in Chapter 9. Now we must draw all the previous intelligence into a coherent whole.

Chapter Seven
# Tying Up Low Weather

The temperate latitude and sub-tropical jetstreams tend to circle the globe and they are inextricably interwoven with the depressions. (Fig. 7.1)

The rules are that new lows form some hundreds of miles on the equatorial side of the jets and their centres then move north-eastwards until they are under the jet. That is when they begin to occlude.

This means that young depressions must have the jetstream to their north which is why we can see jet cirrus ahead

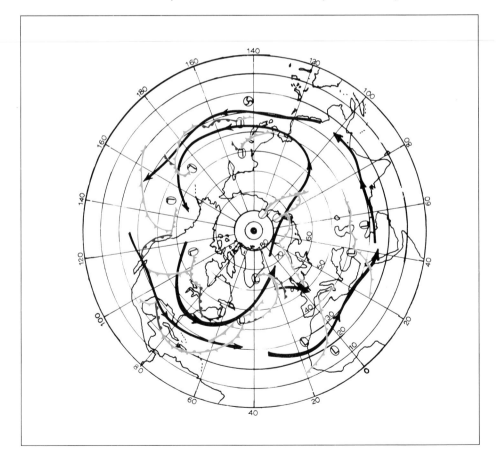

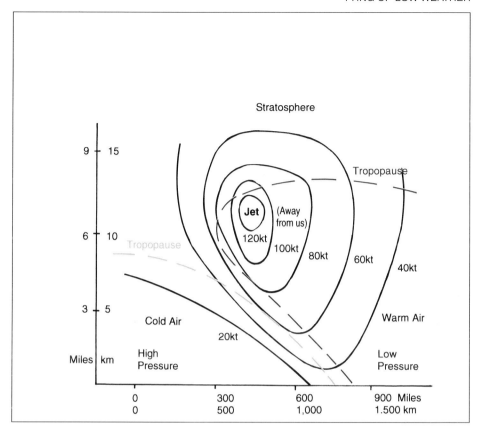

of a newly developing low but not ahead of an old occluded low. There will be cirrus ahead of the occlusion but it will

*⇐ Fig 7.1 Sometimes the temperate latitude jet and the sub-tropical jet form almost continuous girdles around the whole hemisphere. Note that the low at about 50°W is a developing wave depression and is south of the jet. Most of the other lows are occluded and so their centres are north of the jet. The point of occlusion on the 20°W meridian lies under the jet just as it should. There are another two examples at 80°W and 110°W while close to the sub-tropical jet at 140°E is a typhoon in the China Sea.*

*Fig 7.2 Cross-section through a typical jetstream.*

not have the jet form.

The jet cirrus forms on the southern side of the jet not far from its axis but the only places it can be truly seen is well ahead of the warm front and behind the cold front. However, these are the places where we need to be able to see it so that we can make forecasts for the coming hours or even days. After that, other clues will help to make it evident that the foretold weather is actually coming.

In Fig 7.2 we have a cross-section through a typical jetstream and Fig 7.3 shows where the jet will lie in relation to

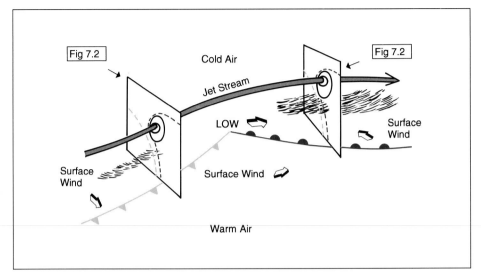

*Fig 7.3 Where the jet and its attendant cirrus lie with respect to the developing depression. The vertical sections are like Fig 7.2.*

a developing surface low. The vertical sections show the frontal surfaces and how the jet lies just in the warm air below the tropopause. We also see the surface wind directions ahead of and behind the low and how they tie in with the direction of the winds at cirrus height.

We can best show where the jet lies over surface depressions with a diagram like Fig 7.4. Here are depicted three lows formed along the polar front. On the left is a new embryonic wave low developing to the south of the jet and travelling fast as waves often do.

In the centre is a low in its prime but just about to start occluding as the centre

almost lies under the jet. Its speed is lower than that of the wave and it has a classic vee-shaped warm sector.

On the right is a low past its prime and

*Fig 7.4 ⇒*
*Three lows along the polar front. On the left a new but developing wave depression travels along the polar front at 40 knots while the mature depression in the centre has slowed down to 30 knots. As the apex of its warm sector is almost under the jet it will soon begin to occlude and start the filling-up process. That is what has happened to the depression on the right. Note how the cloud shield of the mature depression has become a whirl over the occluded one. Typical cross-sections along the line X-X-X are shown below with typical distances which may be widely different from these.*

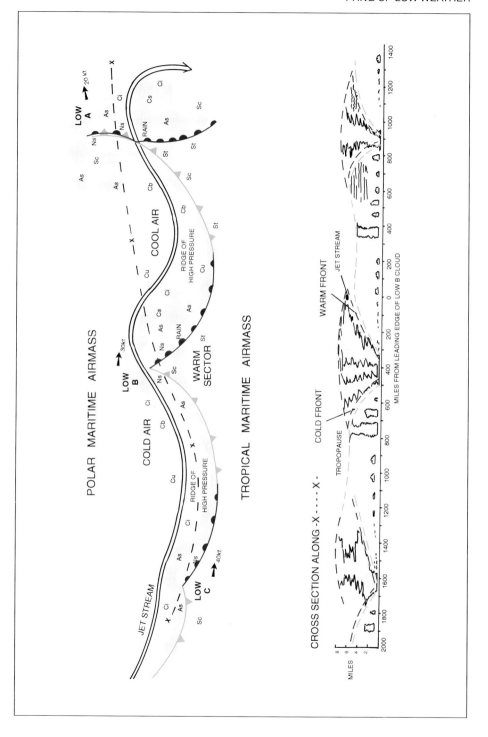

POLAR MARITIME AIRMASS

TROPICAL MARITIME AIRMASS

CROSS SECTION ALONG -X - - - -X -

65

well into the occluding stage of its life cycle. The point of occlusion moves so as to remain under the jet and the low centre is now well to the north. However it sometimes happens that a new low forms on the point of occlusion.

The diagram also indicates the extent of the cloud shields over the low centres and we see that the only shield which looks like a spiral belongs to the middle-aged occluding low. (See photo 3.2). We have also indicated typical places where the different cloud types can be seen.

Figure 7.4 shows a cross-section of the frontal surfaces and cloud wedges along the dashed line X-X together with some of the heap clouds which can form in the ridges etc between the low centres.

The tropopause puts a lid on upward cloud development but note that the 'trop' domes up over the warm air in accordance with the same ideas as we saw with the 300 millibar contour chart (Fig 5.2). Conversely it dips down over the cold air between the lows.

The dashed lines that envelope the different airmasses form corridors along the warm and cold frontal surfaces. This is not just a convention. It is a fact – and many will see it as an amazing fact – that the two airmasses do not mix along the frontal surfaces. As we have only thin air on both sides surely they must run into one-another. That they do not is because they are at different temperatures and so different densities. It is also because there is a third kind of air, dry and originally from the stratosphere, sandwiched between them. This is what is being indicated by the 'corridors' that represent the frontal surfaces.

The clouds of the wave low do not necessarily extend very high but they are building all the time and as soon as they produce rain they must have grown to heights cold enough for the rain-making process to occur.

The maturing low has the greatest cloud masses above its frontal surfaces and so will produce the most rain (or snow) while the occluded front should not produce the same amount. In practice occlusions can rain very heavily for a considerable time because they sometimes get a new lease of life from having a blast of cold air injected into them.

The distances shown are typical ones and with 1,200 miles between the wave low and its mature colleague the wave has room to develop into a full-blown depression. Otherwise, if too close to a low ahead, waves will be stifled in their development and will remain ripples in the polar front, eventually dying out. See the wave south of Low A in Fig 7.5.

## Real situations

We can best illustrate many of the previous text-book ideas by combining the contour chart (Fig 5.2) with its equivalent surface chart (Fig 6.1) so forming Fig 7.5. When we do this we can see the answer to some of the things that may puzzle would-be weather forecasters.

For instance, why is it that sometimes the weather remains much the same for days – even weeks – on end and at other

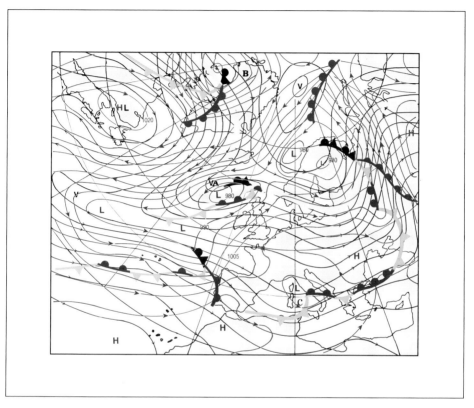

times sweeps through with change after change after change?

Much of the answer lies in the fact that what happens at the surface is controlled by the state of the upper winds. The contour chart for this day in early November has a very good example over Labrador. We have written E for an elevated dome of warm air and V for a valley or trough in the contours. These are not standard letters but we will use them to save ambiguity.

When, as over Labrador, upper E and surface H lie closely over one another

*Fig 7.5 When we combine Figs 5.2 and 6.1 we see that in some places the contours and the isobars cross almost at right angles. These are where the Crossed Winds rules will show the greatest development of the situation either for better or worse. Elsewhere the two sets of lines lie much more parallel to one another and so indicate not much change going on in those regions. Where the upper and lower lows and highs coincide there will be stagnation and the surface weather patterns will not move very far.*

the winds are all going round in one gigantic cylinder of air motion. Such systems cannot move very far and so the surface weather remains much the same type for as long as the upper winds stay gyrating over the surface system. It is a high in this case but it could, when all the winds rotate the other way, be a low.

When a surface low (L) and an upper low (V) coincide (or nearly so) as they do south of Iceland then equally the low will be sluggish. Now we see why the fronts of this low are flying on out of the low itself. The low is being held back by the deep upper trough above it but the fronts are in the 'racetrack' of contours to the south of it. It could well be that a new low will form on the point of occlusion and whirl across into Scandinavia under the impetus of the jetstream above it. Such a new low could be a vigorous one because lows that form in association with jetstreams are lively affairs with strong winds and plenty of precipitation.

There is an upper low (V) over Spitzbergen with a big trough in the contours down into Norway. The lows over northern Scandinavia that more-or-less coincide with this trough will also be slow moving but the fronts over Russia, which were probably the ones with which the lows formed, are going to be moving very fast under the impetus of the jetstream above them.

Then over mid-Atlantic there is another coincidence of surface low (L) and upper low (V), so once again the surface weather is not going to change very much. However to the south is the sub-tropical jet and it is this which is transporting the fronts off Iberia at some speed into the Mediterranean. The low over Italy could well be a nasty one being jollied along by the upper winds to transport itself into the Balkans.

So we have wide racetracks of upper winds where everything is movement and change sandwiched between areas where there is no motion of the weather systems. Whether your weather will be one of constant change or one where the same weather, good or bad, continues for days on end all depends on where the great centres of circulating upper wind happen to rest.

**Waves in the upper westerlies**

It is important to notice that the 'racetrack' from northern Canada down into the Atlantic and thence via Scotland into Scandinavia and so on to Russia forms a series of two giant waves. Similar waves go on right round the hemisphere and an upper-air chart for the Southern Hemisphere will appear very much the same.

You will see that the quasi-stationary high over Labrador is under the crest of one of these waves while the lows over northern Scandinavia are in the trough of another. The lobe-like trough V south-west of Iceland is a bit of an anomaly because it is really part of the great trough in which the lows over Scandinavia have become caught. However it allows us to explain how sometimes the rain in Spain falls in a way which the travel brochures would rather you forgot.

The trough in the upper winds has become so sharp that something has to happen. What happens is that it gets cut off as an upper low while the jet reforms to the north. These upper lows then move into Mediterranean areas and allow the formation of nasty surface lows in association with them.

**Proving the *Crossed Winds* rules**

We have already seen in Chapter 5 that for the weather to develop for better or worse the winds near the surface and those at cirrus height (300mb) have to be crossed to one another.

The combined chart (Fig 7.5) shows some good examples of where the *Crossed Winds* rules operate and also indicate that you can expect to use them anywhere you happen to be outside of the tropics.

Imagine you are on one of the Shetland Islands north of the mainland of Scotland and so in the path of the frontal system that is moving on ahead of low A. Imagine standing with your back to the southerly wind and watching jet cirrus winging in along the edge of the jetstream from a little north of west.

The winds are crossed for a warm airmass on its way but as far as weather goes that means a frontal system and so deterioration in the weather. We shall have the same message if we can compare the motion of cirrus clouds with the flight of cumulus or other low clouds right across into southern Norway. However, across the southern Baltic the

message would be for no immediate change as the upper and lower winds are moving parallel to one another.

This rather uncertain message applies right across Finland and on into northern Russia but then over Siberia and the seas to the north things change. Very strong southerly lower winds are crossed by jet force upper winds. The weather there is going to develop for the worse when the cold air behind the frontal system that stretches back into the Mediterranean moves in from the west.

Turning our attention to Low B over the Arctic Ocean we see strong development for deterioration as the upper and lower winds cross perpendicularly ahead of the warm front and occlusion. Low B is going to move into Greenland because the racetrack of contours show that the upper winds will transport the low rapidly towards Iceland.

A useful analogy is to think of the 'racetrack' as a river and the surface lows as swirls made near its bed. The swirls are transported along by the river but occasionally backwaters appear where the onward flow becomes transformed into slow slack gyrations that are not going anywhere. There are backwaters over Labrador and Scandinavia but the 'river' continues to flow past them carrying the changeable weather with it.

Behind the cold front the message for improvement is not so certain as the upper and lower winds are not so perpendicular to one another. So anyone watching the cirrus behind the retreating cold front of Low B and comparing it with the flight of low coud in the wind not far above their heads would perhaps conclude that any improvement would be delayed as the two would be moving more-or-less parallel to one another. Only when the upper clouds moved from the right of the lower wind could there be hope of real improvement. There is even the chance of temporary deterioration as the little wave runs in over northern Scotland.

Now going to the other extreme of the chart we see that Low C over Italy lies under a less strong but equally determined flight of westerly upper winds. It is going to move into Yugoslavia and deteriorate the weather down into Greece because there are the same *Crossed Winds* ahead of it and behind it as we saw in the model low when these rules were first introduced.

In this game of finding the zones where deterioration or improvement are going to occur we are looking for areas where square boxes are formed between the red surface isobars and the black contours.

A good example is over Biscay. The frontal system which has not yet got a low centre could soon develop one as strong development for the worse is indicated all the way from Spain to Britain but particularly off Biscay. Behind the occluded front there is nothing much happening. There, cirrus and cumulus would be blowing along in much the same direction.

These examples illustrate that wherever you are in the temperate latitudes *Crossed Winds* rules work and we can now re-state them in a way which makes more sense to the would-be weather forecaster who is only interested in whether the weather will get better or worse.

We can now say – and I have proved this time and time again by experience over many years and in many different situations – that the best statement of the *Crossed Winds* rules for the man-in-the-street is as follows: –

### Crossed Winds rule for deterioration

*Stand back to the lower wind and if high clouds come from the left the weather will usually deteriorate.*

### Crossed Winds rule for improvement

*Stand back to the lower wind and if high clouds come from the right the weather will usually improve.*

### Crossed Winds rule for no immediate change

*If the upper and lower winds are more-or-less parallel then there will be no marked tendency to deteriorate or improve.*

If you want to use the rules in the Southern Hemisphere then the only modification is to stand facing the wind.

### The rules in practice

Ever since I left the Meteorological Office many years ago I have used the above rules with considerable success. Having been a forecaster with all the information one could expect I felt naked without my charts but we had been taught a method of gridding the surface and upper-air charts together to forecast what tomorrow's chart would look like. Combined with a set of rules called *Sutcliffe's Rules* we could often make a better stab at the development and movement etc of highs and lows than simply by going on established meteorological principles.

It was through simplifying *Sutcliffe's Rules* for my own use that I came up with the *Crossed Winds* rules and I have found that they work not only on big brash depressions but also on lesser features like troughs of low pressure and slow-moving lows which often defy accurate forecasting by the professionals.

It is particularly when some deterioration in the weather has been forecast and you are waiting for it to arrive that the *Cross Winds* rules come into their own. So often such systems throw out cirrus clouds ahead of them and you can monitor their direction of movement across the sky. They will not usually be going at jet speeds but a TV aerial, the edge of a house, the mast of a yacht in a marina or a boat park etc will always allow you to find the upper wind direction from their motion. If the winds are not truly crossed the weather will be delayed.

Your biggest problem is to find a reliable direction for the lower wind. For reasons we shall come to later the surface wind is not well behaved and does not blow along the isobars. What the isobars give you is the wind that is blowing clear of the frictional drag of the earth which means in practice about

2000ft (600m) up. That is about the base level of cumulus and other low clouds and if you have cumulus in your sky as well as cirrus then it is the direction of motion of these two cloud types which are the most reliable indicators for using the rules.

If there are no low clouds (and medium-level clouds like altostratus and altocumulus will not do) then you will have to look for a flag or smoke in the open or the windvane on a yacht or a clubhouse. After assessing how bedevilled or not it is by obstructions upwind you will then have to rotate it some 30° clockwise if over the land or half that if over the sea to find the true direction of the wind we have called the Lower Wind. (Anti-clockwise in the S.H.)

Windvanes on church steeples or towers or the smoke plumes from tall chimneys are usually reliable and you do not have to worry about what lies to windward of them but you will still have to perform the rotation outlined above.

Assuming you have established a lower wind direction to your satisfaction and have discovered the upper wind direction from cirrus or maybe an aircraft trail you will have then to make up your mind what the relative orientations mean.

As already outlined, perpendicular orientation has the most significance. It means that met forces are working strongly to either deteriorate or improve the coming weather.

If there is about a 45° angle between the two winds then the same message applies but the weather will not deteriorate or improve as markedly as with the perpendicular orientation.

When the winds are almost parallel then it all depends on circumstances what changes, if any, will occur. It has been my experience in summer that when the upper and lower winds blow roughly in the same direction there is often a diminution in the amount of cloud. It can be the exact opposite in winter.

We have now, I hope, convinced would-be forecasters that here is a method of foretelling coming change which will work wherever they may be so long as they do not stray out of the temperate latitudes. They will have to do some work themselves and cannot expect to gain facility with the method all-at-once.

There are going to be many, many situations where you will have to rely on the forecasts to supply the answers. The forecasts have become much more accurate since the development of computer+human forecasting but the forecasters will be the first to admit that with the best will in the world they cannot give you precise times for the arrival and clearance of weather systems if for no other reason than they are restricted by air time. Here is where the layman can do quite a lot for him or herself and the *Crossed Winds* rules will aid that very considerably. We will come back to the rules and their uses on systems like troughs and thundery outbreaks in later chapters. For now we must consider cold fronts and occlusions and what may follow them.

## Chapter Eight

# When Cold Air is Coming

When in the grip of depressions there can be no real improvement in the weather until a more-or-less permanent blast of cold air has become established. Only then can we expect at least a few days of settled weather.

Assume that a family of depressions is producing rain belts, brief warm sectors, cold fronts or occlusions interspersed by ridges of high pressure. These ridges may only last a day or less before the next mass of cirrus is again in the sky, so when can we be sure that it is all over?

We can, of course, consult the forecasts and they will more-often-than-not tell us exactly what is in store – but say we wish to do it for ourselves. To confirm in our own minds that the improvement that is forecast is really going to affect us. Well, firstly we must recognise the attributes of cold fronts.

### The cold front

A cold front separates a warm airmass from a cold one. These terms 'warm' and 'cold' are only relative and while they may have sometimes lost much of their temperature difference near the ground their warm and cold attributes can still be found throughout their depth right up to the tropopause.

Assume you ascend in a balloon in the warm air. A thermometer will tell you that the temperature has at some altitude fallen to zero. We call this the *freezing level*. By human standards 0°C is certainly cold air but compared to the air in the cold airmass at the same level it is warm. The temperature in the latter case may be many degrees C below zero and this kind of difference will extend right up to the tropopause which is what is being illustrated in Figs 7.2, 7.4 etc where the tropopause over the cold air is obviously lower than that over the warm air.

Ana cold fronts will introduce an airmass that is prone to the eruption of cumulonimbus clouds. Such cold fronts are often followed by deep cold airmasses that allow warmed surface air to rise to heights of several miles. The result is anvil-headed mounds of alabaster dotted around the sky with clean blue skies between. The typical showery day that we so often get in spring for example.

We are forewarned of such a cold airmass by the way the upper wind crosses the lower wind – the *Crossed Winds* rule for Improvement as we wrote it down in the last chapter. Now we have to qualify the word 'improvement' because to many people the way a cold front firstly deluges you with squally rain, then, as the rain tails away, clears to blue sky for a while before again deluging you with showers is not an improvement.

Showers that are heavy and cold may seem worse than the weather that went before.

However the *Crossed Winds* rules are giving the observer a means of making predictions about vast blocks of air covering sometimes many thousands of square miles. Sometimes this improvement may not come for a day or two. It is my experience that for a good period of anticyclonic weather to follow a cold front the heap clouds must not pass too rapidly. If they should die out quickly, especially in the middle of the day when they ought to be at their most virile, then take this as a warning to expect another warm front or occlusion in the offing.

Behind a cold front there will usually be a slot of blue sky before cumulus clouds begin to develop. It depends on the time of year how long this takes. You have to realise that as the cloud-edge of the cold front clears over it is progressively opening up cool, moist ground to the sun. After a certain time the ground develops enough thermal energy to start the process of liberating thermals. Obviously that will happen more rapidly in late spring and summer than in autumn and early spring. In winter it may not even happen at all.

The thermals are invisible but the heap clouds that form over them are not. If there is a truly cold airmass above the developing Cu clouds there may be nothing to inhibit the upward motion of the clouds until they reach the tropopause. The result is very heavy showers of rain and/or hail together with some claps of thunder.

You may say 'As a cold airmass always follows a warm one when a cold front passes, why don't we get heavy showers every time?' The answer is:

### Inversions of temperature

It is natural for the temperature of the air to fall as we ascend. We expect it to be colder on top of a mountain than in the valley below, even allowing for the more exposed situation on a mountain. The values given in Part Two for the standard atmosphere called the ICAO Atmosphere show a more-or-less steady fall of temperature as we ascend. The ICAO assumes an average 15°C at the surface that falls to approaching –5°C at 3,000m (10,000ft) and goes below –40°C at 9,000m (30,000ft).

But of course on no day is the real atmosphere anything like the standard and one of the most important ways in which it differs is through layers appearing in it where the temperature actually rises with height (an inversion) or at least remains much the same (an isothermal layer). As their effect on clouds is much the same so we will lump isothermal layers in with inversions.

Inversions are responsible for cloud layers that persist at any level and the decks of relatively warm air that form them can have been put there by frontal action or by air sinking. When air sinks it warms up and this means that clouds are dispersed by sinking air. The 'bones' of old fronts still tend to persist in some of these warm layers. The fronts may long since have ceased to have any cohesive form but layers of altostratus and alto-cumulus clouds often remain or form in the inversion layers.

Another way in which inversions form explains why cumulus clouds do not always grow into cumulonimbus. When depressions move from the Atlantic through the northern seas, great swathes of cold air are drawn down behind them from the polar regions. These cold airmasses come south over steadily increasing sea temperatures and so they become unstable. Myriads of cumulonimbus clouds grow and die out again over the ocean wastes but, as explained under Latent Heat in Part Two, this process transports heat from the sea surface to higher levels so increasing the temperature at altitude. Soon a strong inversion exists which limits the height of the subsequent clouds. When, as in Photo 2.1, the Cu only grows to a limited extent the reason can often be traced to an airstream that once spent a long time travelling south over the ocean and which has therefore developed an inversion at the level of the cloud tops. It is the Returning maritime Polar airmass that is associated with humble Cu clouds and fair weather at most times of year.

The above are ways in which inversions are found well above Earth's surface but the inversions that form near the ground do so in two distinctly different ways.

The first of these is due to the ground cooling with the approach of evening. This inversion forms during most nights when skies are not either wholly overcast or when the wind is blowing strongly. So having an *over-night inversion* is normal. (See Fig 10.1)

What is not so normal is the inversion that is formed by air sinking from aloft in an anticyclone. Because air is gently sub-siding from high up this kind of inversion is called a *subsidence inversion*. We shall come back to this on page 117 but suffice to say that the subsidence inver-sion does not disappear with the day like the over-night inversion does. It sticks for days, sometimes weeks on end, leading to heat waves in summer and 'anticyclo-nic gloom' in winter. In the centres of such anticyclones great dark areas of stratocumulus cloud form and will not disperse. In summer the subsidence inversion concentrates the sun's heat in a relatively thin deck near the ground and this leads to the temperature rising to dizzy heights.

**The over-night inversion**

As the night-time inversion forms most readily when the air is of polar origin so we expect to see it set in strongly after days with heap clouds.

It is common experience that the wind gets up in the forenoon blows most strongly during the afternoon before sinking in speed into the evening.

(Note: forenoon means the four hours from 0800 to midday local sun time as opposed to the morning which may be anytime before that.)

It is no coincidence that the temperatures of the ground and the air

75

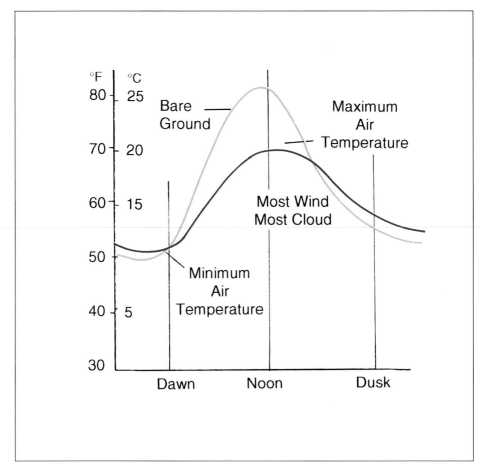

*Fig. 8.1 The diurnal variation in temperature on a summer's day. Note that the bare ground does what you would expect and reaches its maximum temperature at local midday and its minimum at dawn. The air temperature lags on that by an hour or two.*

follow the same pattern nor that the amount of cloud follows suit.

The leader of this meteorological orchestra is the temperature which obeys a curve like Fig 8.1. The cloud amount follows because heap clouds are formed by thermals which do not start to lift off before half way through the forenoon on most days but occur most readily during the afternoon. As the ground begins to cool with the approach of evening so the thermals no longer occur and starved of their source of feed the cumulus clouds die away.

That the wind should sing along to the same tune is not so obvious but understanding why it does so is important if we wish to understand the processes of weather. We shall explain this in the next chapter. In the interim we can state that the effect of the over-night inversion is to put a lid on any attempts by surface air to escape upwards. It also places an invisible barrier between the wind above it and the wind below. The results are:

> Over the land: – the surface wind
> dies away with the night.
> Over the sea: – the surface wind
> speeds up.

The latter effect will be covered in the next chapter where we explain aspects of the wind's behaviour.

### The subsidence inversion

This is far less common than the overnight inversion but nevertheless it has a profound effect on the weather. The last statements above indicate that the night inversion does different things to the wind depending on whether you are on land or sea. The effects of the subsidence inversion are similar wherever you are.

The role of anticyclones is to provide great areas where the air that rises in depressions can sink back towards Earth. When we get blocking highs then this process of sinking from great altitude can go on for a considerable period and so produce a very warm layer of air which may exist within a few thousand feet of the surface.

The air during the day is always going to try to rise if the sun's heat can get to the ground but the subsidence inversion stops it rising. The surface temperature is kept to normal values by warm air rising in thermals and cooler upper air coming down between them to take its place. If this process is inhibited even in the middle of the day and it is cloudless or nearly so all the sun's heat becomes concentrated in the confined space under the inversion rather like an oven with the door closed. Inevitably the temperature rises to record heights.

However another attribute of strong inversions is to concentrate cloud layers and not allow them to disperse. In winter there is very little thermal activity and so thick layers of stratocumulus cloud that are trapped under the inversion have no means of being broken and they persist for days growing thicker and more murky with time. The great 'smogs' of yesteryear were caused by such conditions coupled to a high incidence of coal and other fires. The pollution had no natural means of escape and the pollutants built up to death-dealing proportions as they did in the great London smog of December 1952 when 4,000 people died of bronchitis and other respiratory complaints. Luckily clean air acts have now made such a disaster largely a thing of the past but poor air quality is still to be found and is now forecast on a regular basis.

Chapter Nine
# Low Winds and Weather

The way the wind behaves is a much better weather prophet than a barometer. For one thing you may not have a barometer at home and certainly you have not normally got one with you, except on an offshore yacht. Yet you will always be able to observe the wind wherever you are.

The biggest problem for modern would-be forecasters is how to get a reliable wind direction. As the conurbations grow more and more people become cut-off from the undisturbed weather. For example big towns become 'heat islands' as the combined loss of all that money you are owing the electricity or oil company raises the temperature by some degrees over and above what it will be in the surrounding countryside. In particular they find it more and more difficult to believe that the wind blowing in their garden or the street is a true indication of what the wind would be in the open.

Luckily for sailors who sail offshore this is not such a problem. Out on the sea the wind you feel will be the undisturbed wind but even then you have to allow for the speed of the yacht. The country walker will usually be able to feel the wind that is not too badly affected by trees and buildings. The hill walker or mountain climber may not be in such a good position because the proximity of bluffs and peaks can alter the surface wind out of all recognition.

Thus the problem of finding a true surface wind is quite a thorny one. Yet for most purposes a rough idea of what the wind direction is will do. I, who have always believed in the importance of the wind direction for foretelling the coming weather, have raised a thirty-foot (10m) high wind vane in my garden which, apart from a big sycamore tree to the west of it, gives a reliable indication of the wind's direction.

Not everyone will want to do that, even if they could, so there have to be other things which help to find the wind. In the Preface I described an early experience with the weather. At home we always knew what the wind direction was because the weather cock on the church across the way told us. One day it fell off and was not replaced for years due to lack of money so a large number of the villagers were deprived of a weather prophet at a stroke. You may have a similar reliable indicator of wind direction within sight or may be lucky enough to have an open situation. In any case we will assume for now that you can get a fair idea of what the wind direction is.

## The surface wind
The surface wind does not blow exactly along the direction given by the isobars

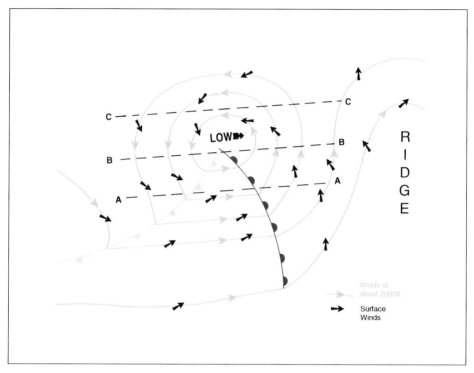

Fig 9.1 To illustrate how the surface winds blow across the isobars out of high pressure and into low. Also the changes in the weather when a typical depression with a warm sector passes.

on a weather map. Isobaric tramlines give you a rough indication but say, as at (A) in Fig 9.1, the isobars indicate a southerly wind. The wind you observe at (A) will not be southerly but more from SE or SSE. This is because of a rule which says:

*Surface winds blow at an angle across the isobars out of high into low.*

How big this angle is depends on the roughness of the terrain. Thus the usually accepted angle is 15° over the open sea and 30° over the land. However sometimes the angle over the land can be less or as much as 45°. (See Fig 12.5)

### Wind shifts and fronts

When depressions approach and pass, the way the wind changes with time follows the more-or-less predictable pattern of cyclonic changes outlined in Chapter 3.

What wind shifts occur depends on where you are with respect to the track of the low centre. In Fig 9.1 assume you are at (A) with a low tracking in as shown. As the sky becomes more and more overcast and the rain begins, the wind will tend to hold this direction until the warm front passes. Then in accordance with the rule:

*Winds veer when fronts pass*

there will be a clockwise shift to around SW. This shift may occur all of a sudden or over a period of time depending on whether the front is an ana one or maybe has been around for some time and has lost some of its original attributes. The latter weak kind of front is called a kata front because the Greek *kata* means down and kata fronts have been weakened by air subsiding over them.

In a well-behaved depression the isobars in the warm sector are straight and so the wind will remain in the same quarter for as long as it takes the sector to pass. In the warm sector the weather may remain overcast and there may be rain or drizzle depending on how wet the air is and how close you are to the low centre. In practice you may find the warm sector wind will back somewhat before the cold front passes.

The cold front has its worst weather at its head and the wind shift that goes with its passage can be sudden and squally. At sea it can result in nasty cross-seas which temporarily make for potentially dangerous conditions. Again it depends on how old the cold front is and a kata cold front may, like the kata warm front, only produce the full wind shift over a period of time.

The veer of wind as the cold front is imminent can sometimes be detected by strange contra motions of the low clouds and such indications usually go with the nastiest conditions.

Now the wind will be somewhere around NW and the clouds will be mainly heap ones while showers are normal but not essential. This phase should last for many hours. If it does not and tall clouds become humbled, while the wind shows signs of backing again, scan the sky for the cirrus of the next depression. Conversely if cirrus appears and increases with time then expect the wind to back later into the south even if the wind has not yet forsaken its westerly quarter. (photo 9.1)

Because the majority of Atlantic lows follow tracks that may start off going east but curve to become north-easterly, so their centres will mainly pass to the north of the major centres of population. Thus the wind shifts just described will be the most likely ones for people in Atlantic Europe and the United States.

However that does not prevent depressions tracking further south on many occasions. Thus you may, as at (B), lie in the path of the low centre. In this case, while you will have the cloud build-up associated with a coming depression, the wind will tend to maintain its direction but instead of increasing with time, as so often happens further south, the wind will slacken and may fall away to next to nothing with low overcast and continuous rain. There is an 'eye of the storm' effect with an unnatural calm descending on everything. You now have to expect a wind from a widely different direction

which may be almost anything until it really settles in from W, NW or even N. Eventually the cloud shield above will break to open sky when you can expect an outbreak of showers but not on every occasion.

When, as at (C), the centre tracks to the South of you the original westerly or southerly wind will back round further east. It is this clue which often tells you that the centre is going to go by to the south. That and the possible sight of cirrus moving from a westerly point while

*Photo 9.1 Cumulus becomes very humble under the influence of high cloud of a front coming in.*

your wind is from an easterly one. Your chances of seeing cirrus will depend on how far from the low centre you happen to be because obviously you have to be out beyond the edges of the cloud shield. Your eventual wind direction is far more likely to be northerly than elsewhere but there will have been no real change of airmass as the low has tracked by.

## Lesser lows

Up to now we have described full-blown depressions which tend to obey the text-book rules. These are the ones which travel and for which the computer-aided forecast system tends to get it most right both in space and time. However not all laws are well-behaved and their fronts certainly play games with the forecasters' best efforts.

The low which comes to a halt with its centre near enough for you to be caught in its cyclonic circulation is one where the forecasts cannot hope to give enough detail on what will happen.

The skies under cyclonically curved isobars will always tend to be overcast and if not actually raining be dark and forbidding enough to make you expect rain. In this general run of poor weather there will usually be minor troughs which will spin round the low centre like the spokes of a wheel. Some indication of their coming can be gained from care-fully observing the wind direction. It will almost always back (shift anti-clockwise) somewhat and there may even be a light-ening of the sky if not a break in the cloud sheet.

It is normal for the wind to increase when rain begins to fall. This follows because the rain drags down faster wind which is blowing up above. If the rain falls very heavily, as it may do when a cold front, an occlusion, or a frontless trough passes, the wind increase is aided by the falling air having to spread side-ways like pouring water onto say a concrete surface. This adds to the squall effect and we are often aware, when it has been cloudy and maybe rain has fallen but not particularly heavily, that when this bluster comes through, the windows suddenly steam up. This tells us at once that a colder airmass has replaced the pre-vious one and so we can now expect a clearance of some kind. Under the aegis of a low's circulation however that clear-ance may be short-lived.

## Wave lows

These have a profound affect on the local weather and wind. They form most readily on cold fronts trailing back from depression centres and they locally deteriorate the weather. If you have kept a careful eye on the weather situation and you are convinced that the rain you are experiencing is from a clearing cold front you will be looking to windward for the bright slot on the horizon that heralds the clearance. You see it coming and look forward to getting rid of umbrellas, macs or oilskins. But disappointment soon follows for the brightness dis-appears and the rain comes on again in earnest.

What has invaded the situation is a wave in the cold front. It can be visual-ised as acting like the hump which you can jerk into a stretched rope one end of which you have attached to something fixed. Just as the hump runs along the rope at high speed so the wave on the cold front will do the same. Also just as the hump does not disturb the rope except where it is, so the wave will only deteriorate the weather in a limited

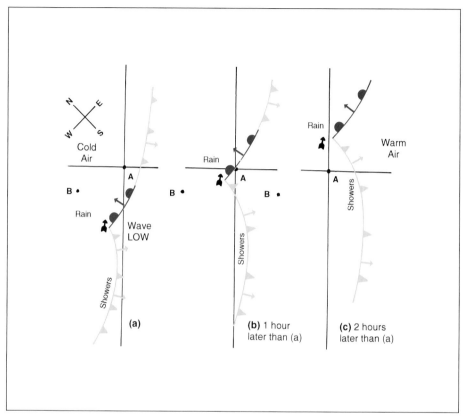

Cold
Air

Rain

Rain

Warm
Air

B •

A

B •

A

B •

A

Showers

Rain

Wave
LOW

Showers

Showers

**(a)**

**(b)** 1 hour
later than (a)

**(c)** 2 hours
later than (a)

locality before it moves on to affect the next locality. Its effect is to slow up the clearance of the front and while it does so it can create several more hours of rain (or snow) than was expected. The modern forecast methods can locate the positions of waves with some accuracy but as they are very mobile, so only you on the spot can see when a wave is affecting you.

The movement of a wave that ran across my home region one summer's day is shown in Fig 9.2. Imagine yourself

*Fig 9.2 The motion of a cold front wave.*

at (A) in Fig 9.2a. the rain has just tailed away but it is still cloudy. Over to the north-west you see the brightness of the clearance. As you watch the cloud closes in again and an hour or so later (b) you are once again under a front but this time it is a warm front as the front is temporarily moving away from the warm air. Elsewhere, although you do not know this, the cold front is moving on. Two hours later (c) you are once again able to see a clearance and this time you can believe it because you only get one wave at a time.

The wind effects are complex when waves pass because due to effects described under Gradient Wind in Chapter 12 the wind tends to blow directly into the tip of the wave as it passes. So the wind might have been NW before the wave but it might even go temporarily east or almost any direction. The consolation is that such a shifty and unpredictable wind is bound to be light.

In this case the wave set off thunderstorms and showers but most waves only lengthen the period of precipitation. In winter, waves on cold fronts can be the reason for locally intense snowfalls. I remember one where in the country we had 18 inches while in the town only nine miles away they had a mere three inches. No one believed me when I rang up to say I could not get in to work because a milk van had effectively blocked the lane by running into a snowdrift.

**Heat lows**
In summer small local lows form or deepen due to the sun's heat. Such heat or *thermal lows* are the catalyst for the outbreak of thunderstorms but they usually form inland and are very rare in coastal areas. They tend to form in the 'centres' of land masses like England and Northern France or Northern Germany. The outbreak of massive thunderstorms needs 'fuel' to be supplied in the form of moist air and thus the storm centres have

to be near enough to the coasts for the sea-breezes to feed them. This is why in Britain the counties to the north of London are the most thunder-prone and that thundery conditions breed over the Ile de France area and often move north from there, while the North German Plain is fed by breezes from both the North Sea and the Ostsee (Baltic). Obviously other regions of the world will experience the same conditions and those who happen to live there will have to contend with some really nasty summer weather.

In the United States it is the warm, moist southerly winds blowing up from the Gulf of Mexico and surrounding seas which make for the central and southern states being so prone to devastating tornadoes. These go with the great thunderstorms of the region. Contrary to popular belief Britain (and Atlantic Europe generally) experience a large number of tornadoes which are often associated with more-than-usually vicious ana cold fronts. They are not so devastating as the US ones and so get little publicity outside of the local press. However to those who lose roofs, sheds and fences etc, as well as to those who experience injury, Europe's tornadoes are not to be laughed at.

**Polar lows**
There is a particularly nasty form of low which spins down from northern seas when the weather map shows an anticyclone stuck out to the west. For Europe that means south of Iceland. These so-

called polar lows do not have fronts like Polar-front lows do but they have the most obnoxious mixture of cold and wet together with almost total overcast. They often produce snow particularly in late spring when we think we have escaped it.

Luckily in Atlantic Europe, it is unusual to have high pressure where low pressure should normally be, and so we do not get all that many polar lows but when we do we know about it.

There are a large number of variations on the text-book low of which only a few have been outlined here but those described are the more prevalent ones.

Chapter Ten

# A Day's Wind

On the majority of days the wind follows a pattern which we have already introduced in Chapter 8. It is called the *diurnal* (*i.e. daily*) *variation* and it shows up most strongly when the weather is anticyclonic and the winds mainly light to moderate.

Situations which upset the diurnal variation include the onset of lows or highs, so that stronger wind is found in say the early morning rather then occurring later in the day. This is opposite to what one would expect because the normal pattern is to have the least wind around dawn and the most in the afternoon.

Then there are the effects of local wind systems like land and sea-breezes and mountain and valley winds – where these can occur. Or the eruption of massive thunderstorms with very strong local winds spilling out of storm cells. These can lead to warm sultry winds of a few knots before the storm, suddenly being swept away by cold 30–40 knot winds from a diametrically opposite direction.

was warmed through to a considerable depth, so an inversion of temperature between the air-deck near the ground and that higher up must occur. This inversion strengthens as the night proceeds and reaches its greatest strength around dawn. The effect of an inversion is to put a lid on the air below it so that the latter stagnates. The wind below the inversion goes down in speed while that above it goes up to compensate.

The inversion sets in late afternoon and so the wind goes down in speed with the evening. It may go up again towards midnight if a nocturnal wind sets in and sometimes when backed by sizeable hills or mountains this over-night wind from the land can get quite strong.

In most cases however the wind will have fallen to a low ebb by dawn and if you are around at that time expect very little wind. If there should be more wind than you expect, then ask yourself why this should be. It usually means the onset of a low which is tightening the gradient of the isobars and so upsetting the normal trend.

**Winds of the early morning**
Let us assume it is a clear or only partly cloudy night. The earth (but not the sea) will cool rapidly over-night because of radiation to space.

Because the air of the previous day

**Winds in the forenoon**
Normally there should not be any great increase in wind speed before breakfast and there should not be any marked change of direction either. The wind should flow smoothly without any great

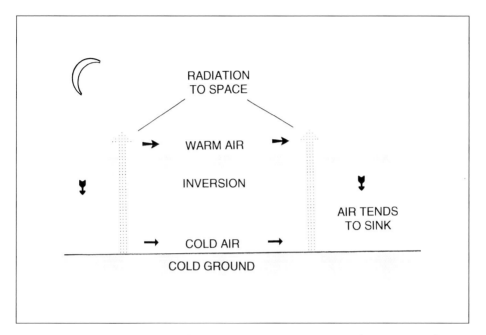

*Fig 10.1  How a surface inversion forms.*

variations in it but soon all that will change.

As the sun gets high enough to be about 30° over the horizon it will, in spring and summer, have enough power to set off thermals. These parcels of warm air will be inhibited from rising too far by the inversion but eventually they will have enough strength to burst through the inversion and break it up. The first signs of this happening are when hill-slopes in the direct sun grow some puffs of cumulus cloud.

The break-up of the inversion allows the stronger wind above to be brought down and mixed with the slow surface wind. The effect is a sudden increase in the wind and often a change of direction as well. The latter is usually a veer. The wind will now strengthen and should continue to do so all the way through the morning and into the early afternoon. The morning wind will however be the most variable of the day and if sailing, you may be treated to some strange direction shifts and sudden puffs. As the wind strengthens towards its afternoon maximum speed so the scale of these variations will decrease. The wind will be stronger but the shifts within it will seem less noticeable.

**Winds in the afternoon**

The couple of hours after when the sun is on the meridian (local noon) will see the wind at its strongest on a normal day.

How variable the afternoon wind becomes depends on its strength and what kind of cloud cover there is.

If, despite the expected increase in speed, the afternoon wind remains sluggish then either you have total low cloud cover or you have no low cloud cover. In the latter case a very strong subsidence inversion is inhibiting the growth of thermals and stopping the air being mixed up through any great depth. If the thermals cannot bring down stronger wind from above to ginger-up the surface wind the latter remains somnolent. There will often be strange shifts of direction as well.

With moderate winds if there should be Cu clouds, the wind will be fairly variable and if these have not grown into anything bigger then it is an inversion near their tops which is preventing them growing. The wind in this case will not do anything odd and will just sink into the evening as the heap clouds disperse.

If the day's wind is strong in the morning then it will usually be quite a bit stronger by the afternoon. This is particularly so if the cloud cover is broken. If it is a total cover of Sc or other layer cloud the increase of speed into the afternoon may not be marked but when the afternoon clouds are heap ones the puffs may be too much for some sailors to handle.

There may not be a sea-breeze when the wind is blowing at moderate strength from land to sea but the opposition of the sea-breeze force to the established wind can make the latter go down with the afternoon rather than up. However, should the day be one of sunshine and heap clouds (or maybe just sunshine) and the wind blows from sea to land the addition of the sea-breeze effect can make the wind in the afternoon exceed the forecast speed by a substantial margin. Then big rollers come onto the beach and children may have to be restrained from being too venturesome.

On such days learners and improvers in dinghies and on boards should go for the morning before the wind really gets up and decide to do something else in the afternoon. They can also try the evening but evening winds are notoriously fickle and not good for beginners.

The afternoon wind will begin to really drop away once the inversion begins to set in. This usually occurs around 16.00–17.00 and then the wind also loses a great deal of its gustiness. Sometimes by the mid evening of a summer's day it will blow as smoothly as the air over an aerofoil in a wind-tunnel. Then it will try to follow the contours of creek and coombe, of valley and hill and some strange local directions will be assumed. Ones which the stronger wind of the day over-rode.

The lower the scale of gustiness in the wind, the more it will be modified to follow the local contours. Winds which blow along with heap clouds tend to be able to over-ride the smaller variations in the terrain and so the wind in hilly areas will not differ much from that on the plains.

Only some of the many variations which can occur in a day's wind have been covered here but they point the way to understanding.

**Winds coastwise**

The most likely reason why daytime coastal winds in spring and summer are not obeying the isobars is sea-breeze activity. The facts about normal sea-breezes which take over the coastal wind system are given in Chapter 12, but winds which blow at moderate or even fresh strength more-or-less parallel to the coast can also be affected.

For example in my own 'home waters' on an east-facing coast, 15–20 knot winds from NE will get a definite cant towards E on a day when it is blue skies and heap clouds. The effect of the sea-breeze force is to act as if the wind were a spring which is pushed in somewhat across the land by day, only to spring back again towards late afternoon.

Winds which may only be forecast to be moderate can become fresh or even strong when off a high coast and the wind is more-or-less parallel to the coast. A forecast moderate wind may become Force 6 to 7.

It cannot be too highly emphasised that winds in harbours and creeks are not at all representative of winds outside. The difference between measured winds in harbour and along the open coast is less by day than by night but it can amount to having to multiply the harbour wind by 1.5 by day and doubling it by night. The wind speeds most affected are those which are 10–15 knots outside but only register maybe 5–8 knots on a masthead anemometer in a marina.

Chapter Eleven
# When Showers Are Forecast

Forecasts of showers have more people giving up on forecasts than almost any other weather element. They may have allowed for rain and nothing has turned up, or the forecast seemed to indicate that there would be showers from the word 'go', but it stayed dry for most of the morning and then a whole crop of showers turned up just before lunch. It can be that you will leave a lowland home base to visit an inland place not more than a dozen miles away and find that while it was dry at home it is raining quite heavily inland. Even more galling is when you drive back home to find they have not had a drop.

That unfortunately is how showers are but that is no reason for not trying to understand their difficult ways.

## What makes a shower?

For showers to occur we need:
1) An airstream which is unstable.
2) Something to start the surface air ascending.
3) Enough moisture in the air.

Unstable air is such that once a parcel of air is set in motion upwards, the air goes on rising of its own volition to a considerable height. That height must be well above the level where the air temperature drops to freezing. Why this is necessary is explained in Chapter 13.

So we need an airstream which grows big solid cumulus clouds. Sometimes what I call 'chimney-cumulus', grows upwards and gets quite tall but just as its name implies it is thin and has very little substance.

Potential shower clouds start growing once thermals can occur. Showery airstreams produce cool clear nights but cool air flowing over ground warmed by the morning sun soon begins to develop thermals. We can see where these, otherwise invisible, thermals are when cumulus clouds top them off.

A thermal source is a local area which gets warmer than its surroundings. Good examples are runways of airfields that are either disused or from which no aircraft are operating, or ploughed fields adjacent to woodland. Anywhere which is undisturbed and cannot absorb the sun's rays in depth will make a thermal source because the surface will rise in temperature above its surroundings. Conversely areas where the rays penetrate but get lost, will stay largely at the same temperature. How a cumulus cloud develops from a thermal source is shown in Fig 11.1.

Air that rises in thermals always cools at 3°C for every thousand feet the air ascends. Our thermal source has been warmed 3°C above its surroundings which are at 12°C.

So when a thermal has risen 2,000ft it

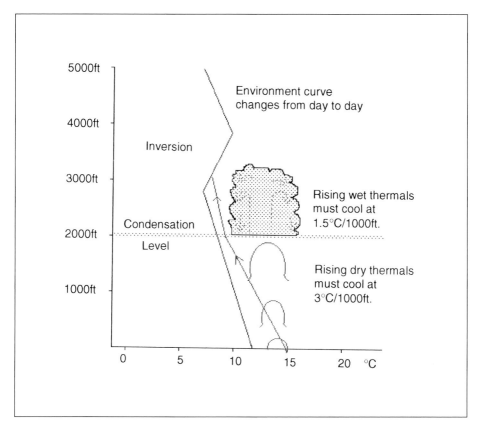

5000ft

Environment curve
changes from day to day

4000ft

Inversion

3000ft

Rising wet thermals
must cool at
1.5°C/1000ft.

Condensation

2000ft

Level

Rising dry thermals
must cool at
3°C/1000ft.

1000ft

0    5    10    15    20   °C

will have cooled by 6°C and so be at (in this case) 9°C. The temperature (the dew point) at which this particular air will begin to condense its moisture and form cloud, happens to be 9°C and this determines where the cloudbases will be.

To form vapour we have to heat water and conversely when vapour condenses into water heat is given back. Thus the air, which is still rising through the developing Cu cloud, now cools more slowly. In fact at half the previous rate.

*Fig 11.1  How cumulus forms and why it does not grow any bigger.*

In another 1,000ft of ascent the air will be a further 1.5°C cooler and, in this example, this is where it has cooled to the same temperature as its surroundings. In this case it does so because of an inversion and the inversion limits the depth of the clouds to a mere 1,000ft.

However the way the surrounding air cools varies from day to day and even from hour to hour. Assume there was no

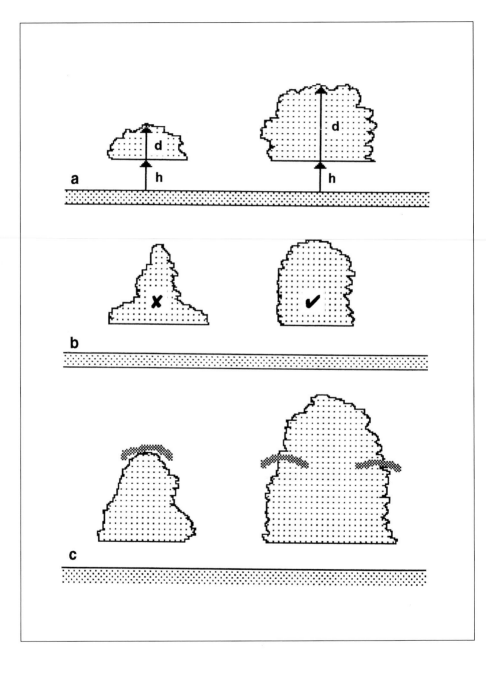

⇐ *Fig 11.2*
*a How to tell fair weather cumulus from the big stuff.*
*b Big cumulus has to be solid. 'Chimney Cu' on the left will not produce showers.*
*c Pileus caps and eyebrows tell us that the Cu is developing into cumulonimbus.*

inversion. So long as the thermal air remains warmer than the environment in which it finds itself it will go on rising. It goes on rising on showery days to heights of maybe 10,000, 20,000 or even 30,000ft.

So an airmass that is cool to great heights is the one which will develop showers i.e. a maritime Polar airmass. In a maritime airmass there is plenty of water-vapour to keep the condensation going as the air thrusts up to these great heights.

What prevents the cloud tops from rising as high as the tropopause is sub-sidence and other inversions. Sinking air over an otherwise showery airstream will form inversions at altitude. These inversions are where the air will stop rising. If the clouds are deep enough for the rain-making process to occur then showers will result. If not then all we will get are big Cu clouds but no rain from them.

**Looking for shower clouds**
A showery day usually starts clear and cool and then Cu begins to develop. The earlier the Cu shows up and the stronger the wind, the more likely it is that the

coming showers will be big and blustery.

When we say the stronger the wind, we have to qualify that because if the wind is too strong, heavy showers do not develop. What is needed is a surface wind that is light to moderate around breakfast-time. The increase with the day will then maybe have the wind blowing at 20–25 knots by the afternoon with maybe 30 knots in the squally showers.

So when the Cu begins to develop and you want to make up your mind whether they will grow into showers later, what you have to look for are the following:
*1. Clouds that are growing deeper than their own base is from the ground (Fig 11.2a)*
*2. Solid trunks so that the tops are as wide as the bases – no 'chimney Cu'. (Fig 11.2b)*
*3. The development of eyebrow-shaped caps of cloud (pileus) over the rapidly growing tops (Fig 11.2c) or left behind on the sides of big clouds which have grown up past them.*

These are a few pointers to the prospect that the fairly humble clouds you have in the morning will grow into something bigger. In particularly if clouds are rushing upwards at such a rate that pileus forms, you are pretty certain to get showers soon.

**The life of a shower**
There is a bit of weather lore that says:
*A sunshiney shower won't last half an hour*
and this is true because the kind of

showers – what we might call 'true showers' – that develop in their own personal spaces, have a life of about half an hour.

All individual cumulus clouds will form in cells and how far one cloud is from another will depend on how high the cloud tops are and whether there is wind or not. What these distances are can be given in a table:

nautical miles (nmi) across and may therefore take three-quarters of an hour to pass, and when thunderstorms break out on sultry summer afternoons, they often do so in cells that are 15–20 nmi across and may only drift very slowly making for maybe some hours of storms. See Chapter 15.

| Cloud top height (1000,s of feet) | 3 | 4 | 6 | 10 | 20 | 30 | 40 |
|---|---|---|---|---|---|---|---|
| No wind | 1.5 | 2 | 3 | 5 | 10 | 16 | 21 |
| Light wind | 1.0 | 1.5 | 2 | 4 | 7.5 | 11 | 15 |

*Table 11.1 Distance in miles between convective cells or streets*

As heap clouds only grow in airstreams where the speed increases with height, so a light surface wind may not be so light in the middle reaches of the clouds. The shower clouds may be moving at 20–30 knots even if the surface wind is 10 knots or less. So if a big shower clouds tops off at 30,000ft and is travelling at 20 knots we see that such a cloud should occupy a personal space of about 11 miles across. That is roughly 10 nautical miles and so at 20 knots would pass in half an hour. So the lore is not far out.

When the cumulonimbus grow to heights where they form anvil heads against the tropopause (Photo 11.1), the personal spaces grow to be some 13–14

**Winds around showers**
The characteristic mode of showery rain is to start heavy and tail away as the shower passes. The heavy rain drags down the upper wind with it and so we expect a bluster of strong wind at the head of the shower. In fact before the shower, very often because the wind blows out round the shower cloud. Big Cb clouds take over the local wind regime producing some backing of the wind ahead of them and a consequent veer when the rain starts. As they pass the wind will often go quite light before picking up again as the next shower approaches.

Showers that are individual, with blue breaks between them, are not the only way we get showers. Often, in inland areas, the showers concentrate in what are called *airmass troughs*. (Photo 11.2).

*Photo 11.1 A big shower approaches. If you are too close to it you may not be able to see how big it really is.*

The airmass trough may look like a cold front and should be treated as one but there will be no change in the airmass after it has gone, unlike a front where the airmasses must be different.

We see in the photo how the masses of Cb cloud are laced together with layer clouds. The air in a trough is rising so it must sink both ahead and behind to compensate. A long period of blue-sky, say late morning when showers are forecast, should make you scan the windward horizon for a coming trough. After the trough there should be another long period without much cloud. Once the sinking phase has passed you may expect more showers. Note in Photo 11.2 how the smoke is showing a very light wind blowing right to left across the face of the coming trough obeying the rule that winds tend to back just ahead of troughs, just as they do ahead of fronts.

⇐ *Photo 11.2 Not to be confused with a cold front, an airmass trough approaches. The forecast charts can tell you that this is not a cold front and so the airmass will not change as it passes.*

## Showers on coasts

Showers occur over the sea when the latter is warm and the airstream is cold compared to it. Thus coasts that face the wind in autumn and early winter are very prone to showers, as are the sea areas to windward of them.

An airstream that blows more-or-less parallel to the main run of a coast will often produce showers on land that sticks out into the sea, without producing any inland. Such coastal regions can receive much snow from which other regions are sheltered.

We have to realise that a shower can only keep going as long as its water-supply lasts. Over the sea showers have an infinite supply of moisture. Once they come ashore they feed on their own reserves and so rain or snow themselves out of existence. Interpose some hills and the coastal plain can experience a 'white-out', while places not far inland, but beyond the hills, see little or no snow.

In spring the sea may be at its lowest temperature but there are also some cold polar or arctic winds at this time. Again showers can occur over the sea. They will certainly occur over the land if the conditions are right.

In quiet weather, nocturnal winds blow from land to sea. With clear skies these winds are cool and they arrive over relatively warm coastal waters. The result is the outbreak of showers over the water and sometimes over the adjacent land when the wind above the nocturnal wind is blowing shorewards. Even if showers do not occur, there will often be a build-up of cloud in the hours before dawn. This cloud (and any showers) will tend to die out as the sun rises.

## Other kinds of showers

The showers we have been talking about above are what I have labelled 'true showers' because they are mainly individual clouds with blue breaks between. However on domestic forecasts it is now normal to label any intermittent rain as 'showers'. This is bad meteorologically as the term 'showers' tells you about the time before the shower and the time after, as well as the time when it is raining on you. When a layer cloud-mass produces rain that starts and stops the indications are very different. Most of such rain comes from altostratus clouds.

Altostratus that produces showers is only partially layer cloud. Go up above it in an aircraft and the upper surface will be peppered with the sprouting cauliflower tops of embedded cumulonimbus. These latter clouds do not have to look like full-grown Cbs with anvil tops. They may only be quite modest in depth but as their bases are already thousands of feet up before they start to grow, they are already deep enough for the rain-making process to occur.

In many frontal systems the rain is not

continuous. It rains moderately, then there are weaker spells or heavier spells and these latter are due to what are called 'mesoscale precipitation areas' (MPAs). Just as described above – and quite invisible to the ground observer – there are showery areas embedded in the cloud-mass (See Fig 6.3) which locally increase the rain for a time.

However this must not be confused with the showers that will follow warm-frontal rain when an occluded front passes. Then – often without a break – the continuous rain of the warm front part, suddenly changes to the showery rain of the cold front part. When that happens you can usually hope for a fairly rapid clearance.

Thundery showers are a product of spring, summer and autumn, and are due to layers of unstable altostratus. Now the upper Cb clouds grow bigger than in MPAs. The rain which falls can be heavy and thunder and lightning are normal. However, you can get showers that you recognise are thundery in nature but with which there is no actual thunder.

Trying to cover all the situations where the forecast will say 'showers' is an impossible task. You have to realise that what you think of as a shower growing great and majestic in a cold polar airstream is not the only entity which will be described as a 'shower'.

## Chapter Twelve
# Facts About Wind

Wind comes about primarily because of motions in the upper air about which we can have little or no knowledge. Increases and decreases and changes of direction in the wind come about mainly due to pressure changes but these pressure changes are themselves the result of two processes called *divergence* and *convergence*.

When a low starts to form on the polar front, the surface pressure falls at some point but that pressure fall is due to divergence occurring in the column standing over the point. What is happening is that at altitude more air is flowing out of the column than is entering it. So the total weight of air is reduced and the pressure it exerts at the surface is lowered. (Fig 12.1)

**Wind and pressure**

As nature abhors a vacuum, so surface air flows in towards the developing low in order to fill it up. It will, in the first few hours of an embryonic low's existence, flow straight down the gradient of pressure, but soon a new factor comes into play.

The air is flowing on a rotating Earth

*Fig. 12.1 Divergence over an embryonic Low.*

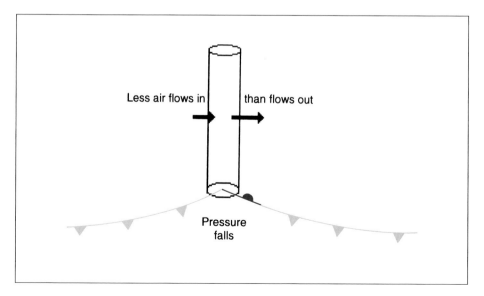

Less air flows in | than flows out

Pressure
falls

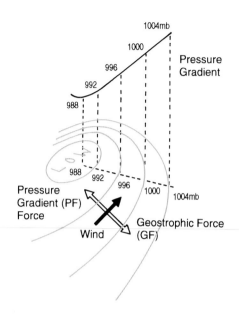

Fig 12.2 *The balance between pressure gradient force and geostrophic force. Above is a vertical representation of what the pressure looks like*

which exerts a force on the air trying to make it move to the right of its path (NH – to the left in SH). This rejoices under the name of the Geostrophic Force (GF). However the pressure gradient force (PF) is not altered by this and eventually the two forces become equal and opposite while the wind blows perpendicular to both of them. (Fig 12.2)

The size of the GF depends on how fast the wind blows but it has to adjust to be the same size as the PF. Thus if the pressure gradient is steeper the GF and so the wind speed has to increase to compensate.

This leads to the well known fact that:
*Strong winds go with closely spaced isobars*

We speak of the pressure gradient (or just the gradient) 'tightening' when the pressure is falling in a deepening low and the high pressure close by is not altering much. Or, as in Fig 12.3, a low is travelling past a high pressure region. The observer at A is enjoying light winds in a ridge from High A but the coming Low D will soon tighten the gradient unless High A gives way – which it often does not.

The observer at B is experiencing the strong winds which go with the tight gradient between High B and Low D. Here is where gales will occur and if, as shown, the position B is under cyclonicly curved isobars there will be much cloud, rain or showers.

The strong wind situation will last until the low has tracked by, after which the gradient will slacken again as the next ridge, this time from High C, exerts itself. However not everywhere may enjoy light winds because, as at C, the local pressure gradient may remain tight before eventually slackening. But as the isobars gain an anticyclonic curvature so the amount of cloud and the risk of precipitation are both reduced.

High B shows an example of how the strong gale can rapidly blow up because if High B's pressure is rising rapidly while Low D's is falling, the gradient will tighten very quickly. Here is where the barometer may fall by more than the magic number of 3 millibars per hour

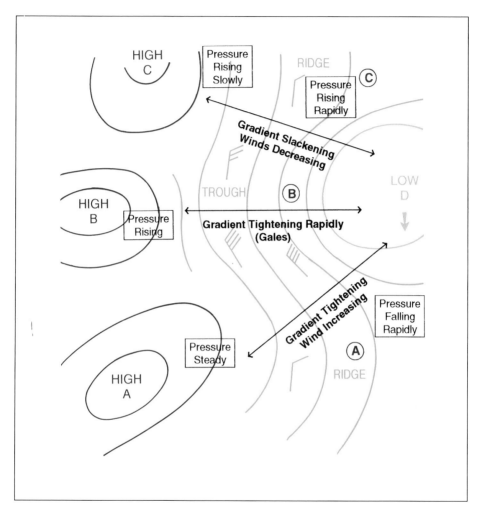

Fig 12.3 What tightens and slackens the gradient as a Low tracks past surrounding highs.

which triggers an automatic gale warning.

Another situation which induces very strong winds is when a secondary low develops between a low like D and the high such as B. The westerly winds on the southern side of the secondary can grow to Force 10 in these conditions, while not far away the easterlies on the north side of the secondary can be light or moderate. However if the winds do lighten in this fashion, when it was

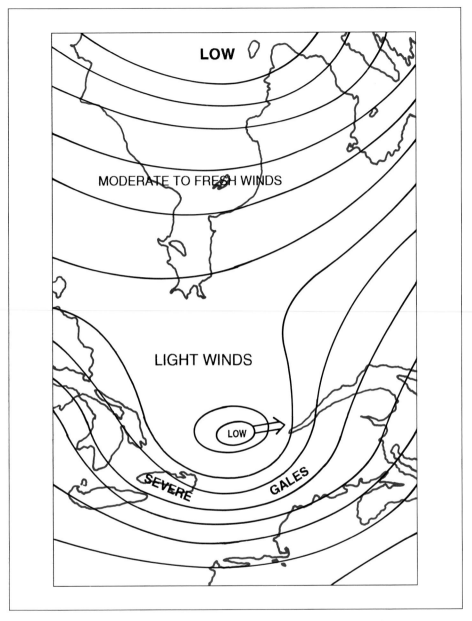

*Fig 12.4 How a little Low that runs into the circulation of a bigger one can produce local gale-force winds.*

expected that they would be strong, do not trust the light phase for long. It will probably be swept away by strong W or NW winds later.

Here is an example of a small secondary low that interposed itself in the circulation of a large one. The main low is up over the north of Hudson Bay but the fact that a subsidiary low would develop in its circulation could have been picked up by the computer-aided forecast as much as two days or more ahead. So the severe gales near to and over the Lakes and the coasts of Massachusetts and Maine would, we assume, have been forecast well in advance. These small lows tend to run on fast carrying their weather with them and we must expect winds, maybe up to storm force, over the Gulf of St Lawrence soon. The St Lawrence Seaway will escape the worst and it is interesting to realise that to the north, within a hundred miles or so of where it is blowing very hard indeed, the winds will be light or even calm. (Fig 12.4)

You can test if there is a low moving past between the primary low and the high when you can get a sight of cirrus and which way it is moving. It should be going in a diametrically opposite direction to the surface wind. (See p 70)

The clues from the sky outlined in earlier chapters must trigger the response of seeking a forecast to see what they say about the wind in your area of interest. The computer-aided forecasts are pretty accurate when it comes to forecasting strong winds although they are more liable to error when the wind lightens.

**Wind and coasts**

The friction between wind and sea is much less than between wind and land, however smooth the land may appear. This means that if the wind is blowing from land to sea it has to increase in speed once it has shaken off the friction of the coastlands. As a rule we can say that:

*If the wind speed over the land is V kts it will be 2V over the sea and 3V up above the friction layer.*

The wind speed given by the isobars is the one which is free of the Earth's friction, so put another way, if the wind given by the tightness of the isobars is say W kts, then the wind speed below over the land is a third of W while that over the sea is two-thirds. (Fig 12.5)

On a day when cumulus clouds are sailing out to sea at say, 30 knots, we can expect the wind over the land to be 10 knots on average, while that over the sea is 20 knots. Thus the wind has to gather speed as it leaves a coastline and at the same time it will veer by an average 15°.

When the wind blows from sea to land it will come ashore say at 20 knots but the speed will soon fall to 10 knots as the air moves inland.

When the wind is blowing more-or-less parallel to a coastline the wind over the sea will be travelling faster than that over the adjacent land. The effect is to make the sea wind tend to invade the beachlands.

However the effects at night are more dramatic than these averages would indicate. The wind over the land goes down very markedly by night and may almost go calm, even though the wind above the night-time inversion is blowing as fast as say 20 or more knots. There is often no inversion over the sea, except when the wind is already blowing from a

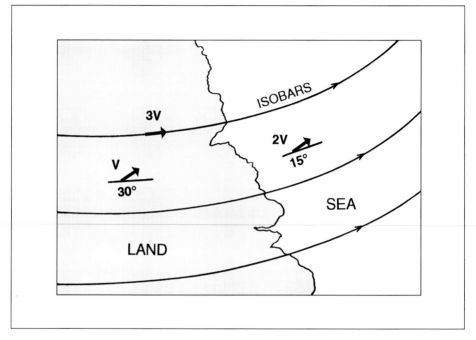

*Fig 12.5 Wind speed and direction over land, sea and at isobar height.*

sizeable land area upwind. Thus the wind over the sea will be more turbulent and stronger than that over the land.

It is certainly the case that when the evening wind does blow along a coast, the wind over the land will fall in speed, while that over the sea goes up to compensate. Just as when the land wind falls with the onset of evening, the wind above the inversion, that is causing the fall, increases to compensate.

These effects taken with the complications due to coastal indentations and promontories make the coastwise wind one that is bound to differ considerably from that given in land area or sea area forecasts. Only on-the-spot and with a good look at the chart can one expect to divine exactly what the wind direction and speed will be. Actually a yachtsman might be laughed at for taking a contour map to sea with him but it could pay dividends by revealing where the coastal land rises and falls, so indicating where the wind may be given unusual cants.

It cannot be too strongly emphasised that a wind from an oceanic direction will not be subject to the diurnal variation experienced on land. In fact the night wind over the open sea will normally increase even though there is no change in the local pressure pattern.

## Local winds

In temperate latitudes we can fairly safely assume that a wind of 15 knots or more (Force 4 or more) is due to the local pressure pattern and that the wind will be blowing at the behest of the isobars. If the wind is less than this we cannot be sure that it is not a local wind. Certainly this should be true from evening right through the night until mid-morning. There will be places, backed by mountains or high hills, where the air draining off the land by night will make nonsense of the above rule, but given normal conditions it is quite a useful rule. Time of day will often enable you to make up your mind that the wind is not a local one such as a sea-breeze.

## Sea-breezes

These are winds that blow from sea to land during the day and for most temperate latitude coasts a wind of 10 knots or more at breakfast-time will preclude them starting.

### Sea-breezes need:

*Light winds*

*Fair morning skies so that the sun can get to the land*

*Limited instability so that cumulus develops.*

### Sea-breezes will not develop when:

*The morning is overcast*

*The wind is blowing from land to sea at above 10 kts*

*It is a hot and sultry morning.*

Occasionally breezes do develop when a too-strong morning wind has decreased through the morning, allowing the breeze to invade the land during the afternoon. Such breezes do not extend very far inland.

On the other hand, given the right conditions, breezes can get fifty or more miles inland during late spring and early summer.

What happens on a sea-breeze day depends on what the morning wind is doing. If it is blowing on-shore (i.e. from sea to land) then the sea-breeze just serves to increase the existing wind and maybe alter its direction somewhat to come more directly on-shore.

The more important case is when the sea-breeze has to first of all nullify and then reverse an already off-shore (i.e. from land to sea) wind. The lighter the wind, the earlier a sea-breeze can assert itself. For example when the morning is flat calm, sea-breezes can get to places three miles inland by as early as 08.00.

As the wind speed opposing the breeze increases so it becomes progressively more difficult for the breeze to stop and reverse the wind. Morning winds of just 7–10 knots usually mean no breeze before the afternoon.

The breeze and the wind fight their battle just off the coast and typically the breeze has reduced the coastal waters to calm by about midday or before. Then the breeze invades the shore and a mini-cold front, called a sea-breeze front develops between breeze and wind.

The situation depicted in Fig 12.6 could be about lunchtime on a coast that is fairly flat and faces between East and South. Backing hills or mountains will

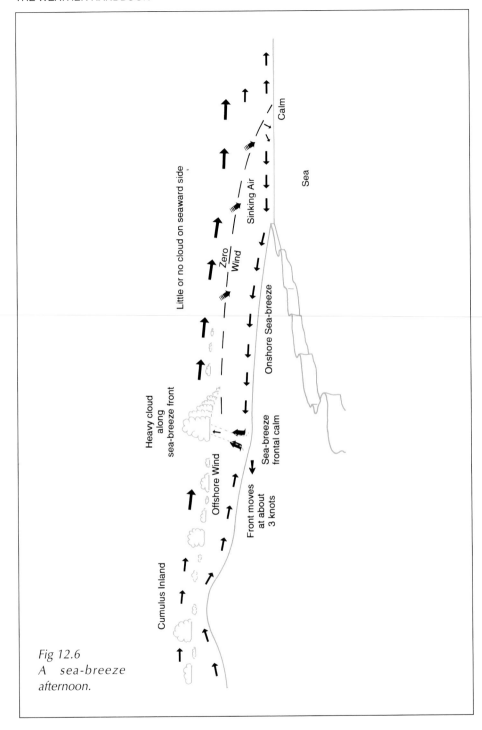

*Fig 12.6
A sea-breeze
afternoon.*

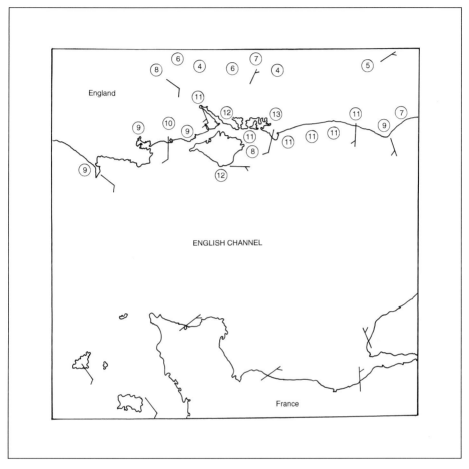

*Fig 12.7 Sea-breezes in the English Channel and hours of sunshine on the coast and inland.*

thwart the breeze's progress inland although they may aid it starting by reducing the wind speed over the coast and by breaking up cloud layers that cover regions further inland.

Coasts that face west or north see the morning sun less well than those that face south or east but they will keep the breeze going longer into the evening as the evening sun will not desert them as it will the south to east coasts.

The sea-breeze front is revealed by a 'line' of more-than-usually developed Cu clouds which lays roughly parallel to the general run of the coastline. It is in a 'chimney' up which the two winds blowing into one another find an escape. On the seaward side the air will be clean and cool and there will most often be clear blue skies. On the landward side there will be much more Cu cloud and a more hazy atmosphere that one generally finds over the land. We can see the effects of this in action from Fig 12.7. The figures in circles give the hours of sunshine on this particular early summer day and we see how the coastal resorts enjoyed many more hours of unbroken sunshine than places inland. This comes about because the air on the seaward side of the sea-breeze front has to sink and in so-doing it rubs out the Cu clouds there leading to total sunshine.

There are other advantages for those who find themselves within the throw of the sea-breeze. People who are prone to air-borne allergies will find the pollen-count much reduced when the sea-breeze blows in. Air from the sea will usually have a much lower pollen count than that from the land.

Detecting that the sea-breeze is imminent is often not a difficult task because there is a calm area below the sea-breeze front. Thus if the wind goes down in late morning or early afternoon when you know it should, by rights, be going up, the sea-breeze cannot be long delayed. In the gardens of coastal hinterlands this can well be the hottest part of the day before the breeze brings coolness for the afternoon.

Sailors on small-water arms of the sea and rivers etc must look for the line of more massive Cu clouds that make up the visible signs of the front, as well as noting the falling off of wind speed and another useful pointer – the sight of soaring birds wheeling in the updraughts of the front. Here is where we can see the strange sight of craft goose-winging towards each other in opposite directions. They will not collide head-on because they will both run out of wind under the sea-breeze front.

**The sea-breeze season**
It is the difference between land and sea temperatures which forms the sea-breeze current and so the breeze will be strongest and most prevalent when this difference is greatest.

Sea temperature will, on most coasts, be at their minimum in February and at their maximum in September. The sea is a great storage heater which takes a long time to warm up but once it has warmed up it also loses its heat slowly. The land on the other hand rapidly warms up and cools down with the rhythm of day and night.

So late spring and early summer sees the greatest daytime difference between land and sea and so the greatest number of sea-breezes. It is at this time that a breeze that starts from the coast at say 10.00 (LST = Local Sun Time) can still be found slowly rolling inland as late as 22.00. The air is still moving forward under its own momentum even though the sun has long since failed to have any influence on it.

The breeze blows on many days – in good months on half the days of the month – right through the summer but begins to weaken considerably as late summer gives way to autumn. Now the land is cooling while the sea has risen to its highest temperature and so very little difference appears between land and sea. Only on very light wind days will you get sea-breezes in autumn. There are a few when it is otherwise calm or light airs under blue skies in winter but they will only be felt by those on the coastline itself.

## Nocturnal winds

Winds that blow off the land at night need very quiet conditions. However the over-night inversion will often, in anti-cyclonic weather, lead to calm during the evening and first part of the night.

The nocturnal wind is a mixture of *land breeze* and *katabatic wind.*

The land breeze will blow from land to sea whenever the night is cool enough for the temperature to fall below that of the sea. It can do so on a completely flat coastal plain as for example in the Low Countries. It is the night-time equivalent of the daytime sea-breeze but is nothing like as strong on most coasts.

Only when the coast is backed by hills will the land breeze get help from cool air sinking off the hill slopes towards the sea. This down-slope wind is the katabatic and it can make the nocturnal wind something to be reckoned with.

On some nights the nocturnal wind is the only wind that a coastwise passage-maker will find and it will be found close in to the coastline. Further offshore it

sinks to nothing but it can become 10 or more knots when funnelled out through harbour mouths or coastal valleys.

The nocturnal wind blows best in autumn when the land is cooling rapidly while the sea still retains much of the heat it stored during the summer. Spring, when the nights are still cold and the sea is cold will induce no land-breeze but there can be katabatics. Anywhere at the foot of a mountain valley or similar can expect strong katabatic-type winds during the night.

## Mountain winds

In mountainous districts it is rare to get a surface wind which obeys the rules and blows more-or-less along the isobars. The effects of peaks and valleys are always being felt and only those who live in a certain locality will know fully about their local winds – and what their names are. For the local winds of regions like the Alpine foreland and the Rockies gain special names of which Föhn and Chinook are simply the most well known.

Föhn winds are dry falling winds which come when a southerly wind from the Mediterranean tries to scale the barrier of the Alps and in so doing loses much of its moisture on the Mediterranean side. It eventually bursts through the gaps and comes down the valleys sometimes at gale force but with an unnatural dryness.

Anyone going to the region for the first time should take care to consult the locals about the effects of Föhn and whether any is expected. Possibly as much as some days before the Föhn arrives it exerts a psychological influence on people in the Foreland area making for increased risk of accidents etc.

Similarly the Chinook is the Föhn wind of the Rockies producing much the same effects although of the two the Chinook can be the more devastating.

Anywhere that wind is thwarted by a substantial mountain barrier can have Föhn-like effects. The Scottish east coast has an exceptionally mild climate due to the warmth of the winds that descend the slopes of the Scottish Highlands while the English north-west coast experiences a noticeable warming when NE winds blow over the Pennines.

Another form of mountain gap wind is the Mistral (and its Adriatic equivalent – the Bora). Mistral needs a situation with low pressure over the Balkans and high pressure over Biscay. Then the cool northerly winds will fall through the valleys (especially the Rhône Valley) to sweep out into the Gulf of Lions and as far east as Genoa, attaining their maximum strength where the valleys constrict them but still rising to gale force on many occasions over the coast. Only in October and November can you expect to be free of gale Force 8 Mistral between Perpignan and Marseille but in the spring

up to half the days of a month can have Mistral of Force 6 and above. Similar remarks apply to the north-easterly Bora on the north-east coasts of the Adriatic.

Again these are the mountain winds that are used as examples of such winds. There will be similar, but lesser, effects wherever winds blow through mountain gaps and over ridges. Those in lee will find they are warmer than elsewhere as the air compresses and so warms on its downward sweeps.

## Inland local winds

When the large-scale winds are light many local influences can induce odd winds. One of these is the creation of heat (thermal) lows inland on warm days. Winds will flow straight into such lows when they first form but after a few hours they will show some form of veering to allow for the Earth's rotation.

With calm morning conditions and blue summer skies local hot spots can form and draw in light local winds for which there is apparently no reason. Bodies of water can form the source regions for such winds particularly where the water lies within sunlit slopes. There will be anabatic (upslope) winds from water to slopes as well as katabatics from slopes in shadow.

Chapter Thirteen
# Rain and Snow

In this chapter we are going to explain how things like rain, snow, drizzle, hail etc form and under what circumstances.

## Drizzle

Drizzle is not rain. If it were, drizzle would not have a special name. The two are only alike in being water drops but their mode of formation is very different. Technically it is drizzle if the droplets are less than half a millimetre across but that is academic. No one is going to start measuring drizzle drops to see if they are technically rain.

We know when it drizzles and are not fooled into calling it rain. Drizzle is very much more dense than rain and it seems to saturate you quicker than you expect. It comes out of low layer clouds and it goes with poor visibility and relatively warm air. It does not normally drizzle when it is cold and we associate it with the warm sectors of depressions or the much more persistent SW wind regimes that steam up from southern seas. It often drizzles for no apparent reason in these airstreams but the cause can often be traced to rises in the ground that induce the wet air to locally rise.

When that happens minute cloud droplets combine together (coalesce) into larger drizzle droplets and so become heavy enough to fall out of the cloud. This process of coalescence is what we call a 'warm cloud process'. However coalescence can never produce deluges of rain – that needs another mechanism.

## Rain

Most rain starts off as snow. If it rains hard on a warm day that rain started life as snowflakes. This is why rain clouds have to be deep or at least high enough up to be cold enough for snow. In fact much colder than we would associate with snow.

It is virtually impossible for water molecules to fuse together to form the nucleus of a raindrop. So some other process must occur.

If we go up to where the temperature has fallen to below $-13°C$ many of the particles forming clouds are ice crystals. Ice crystals are six-sided because of the way $H_2O$ molecules join together and water molecules have a natural affinity with this shape and latch on to the ice crystal nuclei in prodigious numbers. The wetter the air the faster the ice crystal nuclei grow and eventually large snowflakes form. These come slowly down through the cloud melting into raindrops which warm up as they fall to Earth.

The birth of raindrops as snow flakes helps to prevent the drops from breaking into many smaller drops for the flakes act as a parachute and so lower the potential rain drop to heights from which it can fall without friction disrupting it.

Obviously the higher the freezing level (zero isotherm) the warmer the rain can be. On the other hand, when the freezing level is close to the ground, as it can be in winter and spring, the snow flakes will not melt into rain or they may do so partially and so we get sleet.

This theory of rain formation explains why only deep clouds produce copious amounts of rain. It is possible to obtain small amounts of rain from warm clouds but almost all our rain starts off high up in nimbostratus or cumulonimbus clouds. It explains why most of our continuous rain comes from the deep cloud layers of fronts and why, when we are wondering if a mass of heap clouds bearing down on us will break out into showers, we must always be looking for the tall cloud tops amongst the rest.

### The windscreen-wiper rain-gauge

Driving a car when it is raining tells you much about what kind of weather you are experiencing and what is likely to happen later. Here are a few ideas:

### Intermittent sweep

*(1) An old front or maybe the start of the rain of a more vigorous front. The latter often produces a short shower before the rain starts in earnest.*

*(2) Could be the last rain of a cold front before it breaks out into clearing skies. (Previous history will make this apparent as an option).*

*3) If drizzle is this light then it may be a very old front indeed and not likely to produce much else.*

### Continuous sweep

*(1) An ana warm front or an occlusion.*

*(2) Sometimes this kind of rain falls from a frontless trough.*

*(3) The weather may be showery but if so you will know by its stops and starts.*

*(4) The rain of a cold front just after the heaviest downpour has ceased.*

*(5) If drizzle – maybe a warm front has just passed or you are going to the coast or into the hills with a wet, warm airstream.*

### High speed sweep

*(1) A cold front is passing or the cold front part of an occlusion. This will often reveal itself when you also need the*

*demister to clear the condensation that suddenly occurs.*

*(2) Cumulonimbus – more heavy showers to come, maybe a thunderstorm.*

*(3) If drizzle you are probably in a hilly district with the wind blowing up the slopes. If you can still see now, allow for thick fog soon.*

## Rain and driving

Once upon a time weather came to people because they had no means of going swiftly to anywhere else. Now we all travel into other people's weather.

Drive at 70 mph westwards on a motorway and your relative speed of passing the average front is 100 mph. So the weather clears in a third of the time it would were you static. You also run into the heaviest rain much more rapidly than you would expect. If it has been really heavy you should always keep an eye out for floods, especially when you turn off the motorways into the side roads.

Driving like this into a sharp cold front will make the rate of collection of water so great as to force a slow down because the best of wipers cannot cope. The same remarks apply to heavy showers but in the frontal case there is all the cold front weather to come whereas with showers you should soon be out of this one – and into the next one!

Going with the weather – usually from west to east – means that you stay with it that much longer. Assuming the average front moves at 30 mph then your over-taking speed even on a motorway is only about the same figure. Thus the rain scheduled to last say four hours will still last four hours. It will be a very wet and nerve-jangling journey and any adjustment you can make to your journey time to miss the worst will be worth it.

I well remember two separate holidays in the west when we drove back east on the slow holiday routes tailing cold fronts. All the time we had to contend with flooded roads but the sun shone as the back edge of the cold frontal cloud moved eastwards at about the same average speed that we were able to make. That was some years ago when average speeds were lower. Today we would probably have had wet journeys on both occasions as we slowly overtook the fronts.

A similar situation arises with showers. Showers inject a great deal of surface water onto the roads and make driving hazardous enough, but if your route should carry you in the same direction as passing showers the shower may last for a long time. On the other hand driving into the wind will have you through a shower very quickly but make your contact with the next one that much more rapid.

## Snow

As already pointed out, when the freezing level gets down on the ground rain is going to be snow.

There is a psychological factor in forecasts of snow. Rain goes down the drains but snow lays about for all to see. So forecasters, who are very aware of the disruption which is caused by snowfall, will tend to 'over-forecast' so that often threatened snow does not materialise. Or if it does it may well come later than forecast.

There was a good example recently when the morning forecast went for a band of snow coming down from the north-west in association with a small frontless low. It was problematic whether the snow would cross us or just clip us as it moved down towards the south.

There was quite a mass of cirrus at breakfast-time that was moving quite briskly from NW and as the windvane also showed a NW wind, the *Crossed Winds* rules indicated that any onset of the snow would be later rather than sooner.

I continued to monitor the wind during the day as the cloud increased, sometimes with a wan sun peeking through hazy holes in the cloud build-up. By mid-morning the wind had backed W and by afternoon had gone to SW. As upper winds only change slowly so I continued to assume that it was NW up there. The *Crossed Winds* forecast was now definitely for deterioration and the lunchtime forecast confirmed that the snow area was more extensive than originally thought and we were certainly going to get it.

This forecast went for some deepening of the low which is often a sign that it will slow up – not having enough energy to deepen and travel at the same time. This meant that we had to wait until after tea before the snow started. When it did so it

was very fine stuff and covered everything in a fairy dusting that sparkled in the street lamps. The low had lost a great deal of its snow earlier in its path, so we did not, on this occasion, get very much. However as there was already some fourteen inches on top of the cars we did not feel we wanted much more.

The temperature plummeted behind the retreating low and next morning was highly treacherous with ice-covered pavements and roads plus the additional hazard of freezing fog.

This is a good example of the help that the *Crossed Winds* rules can give you when the timing of a coming weather feature is in doubt. It so often happens that lows and fronts do not move on across a land mass at quite the speed they crossed the coastline especially when a mountain range stands in the way as it did here.

Heavy snow cannot occur if the air is too cold. The snow that falls when the surface temperature is well below freezing is usually only a dusting and is in the form of grains or pellets – often called 'sago' snow. In an already cold situation where snow is possible any rise in temperature should be taken as a warning of coming heavy snow. It is practically impossible for moderate or heavy snowfall to occur unless the temperature has risen to a couple of degrees or so below freezing.

A great problem for the forecasters is when the conditions are marginal between rain and snow. The smallest changes in temperature distribution near

the ground can make for snow in one place, sleet next door and rain where maybe the air gets a slant off the sea. You will find a certain vagueness creeping into the forecasts on these occasions.

Care is needed when the freezing level lies within a few hundred feet of the ground. Climbing to say 1,000 feet is not very unusual but if you do, then what is rain on the lowlands will become snow at a moderate altitude.

Chapter Fourteen
# High Weather

In Atlantic Europe there always seem to be more periods of low weather than of high. This is because much of Europe lies in the path of the polar front depression factory but during the summer months a ridge from the semi-permanent Azores anticyclone does push up across Biscay and Britain and as far north as Scandinavia.

That is what happens in the average summer but recent weather patterns show signs of deviating from the norm quite considerably. Thus up to 1992 there had been several years of below average rainfall in Britain. Such a period of drought must have more than its fair share of anticyclonic weather because the process that creates anticyclones goes with low precipitation.

Anticyclones are regions where the air sinks from very high altitude. When it reaches the ground it has to spread out sideways. Thus highs are regions of diverging winds while lows are regions of converging winds. Air cannot move in on itself without going somewhere and that is upwards, but the way this happens is not simple.

Until relatively recently how anticyclones fed depressions was not understood. Then Doppler radar studies proved that the anticyclones that stand to the south and south-east of a major low fed the latter with their air up two great ascending 'rivers' called conveyer belts (Fig 14.1). The high ahead of the low tends to have cold air in it and this forms the source for the 'cold conveyer belt' which travels up the warm front to eventually turn and join the upper westerlies.

The high to the south tends to have warm air at low levels and this feeds the 'warm conveyer belt' that climbs up over the warm sector and again turns to join the upper westerlies but at cirrus levels. So this theory does not obviate our *Crossed Winds* rule for coming warm air and so bad weather. The cold dry airstream that comes in behind the cold front will account for the clear slot that almost always appears behind the front's cloud edge.

So highs which are traditionally associated with good weather are also the source of our bad weather and also a good example of:

> *What goes up must come*
> *down*

While we may not get as many highs as we may like, they must be somewhere not far away to redress the balance when lows transport all this air aloft.

### Blocking highs

Highs that can feed travelling lows must also be travelling. They are anticyclonic whirls in the bed of the westerly stream

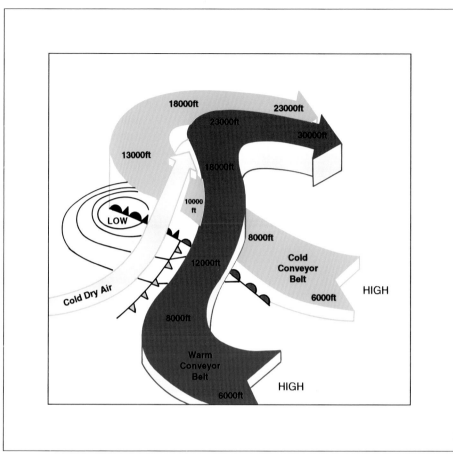

*Fig 14.1 The modern view of how surrounding highs feed a low up 'conveyor belts'.*

and have to go along with the main flow. However the great waves in the upper westerlies (Chapter 7) sometimes become so distorted that they stop moving round the hemisphere and where they form great stationary wave crests anticyclones nestle in the waves and do not have any tendency to move anywhere. Such quasi-stationary anticyclones that are cosseted by the upper flow are called blocking highs.

One of the most extensive blocking highs in living memory covered the whole of the North Atlantic area during the summer of 1975. (Fig 14.2) How the block stands four-square in the path of the lows and diverts them north and south of it is evident in the figure. In such cases places that normally experience excellent summer weather such as Spain and the Mediterranean generally are fed with unaccustomed lows and we also find lows invading the territory of the polar high.

Anticyclones tend to two distinctly different kinds of sky. There is either little or no cloud or there is a total cover.

The clear skies are produced by the

warming air that is compressed as it sinks. This leads to the evaporation of clouds and means that the upper sky is going to be largely cloud-free. If the lower levels are also cloud-free then we have untrammelled sunshine. However

*Fig 14.2 An immense blocking high diverting the lows during the long, hot and dry European summer of 1975.*

the sinking air also creates inversions not far from the ground in which dark impenetrable masses of stratocumulus cloud form. Even in summer the sun may not have the power to burn-off this cloud and so it remains permanently cloudy. In winter the cloud layer leads to 'anti-cyclonic gloom' and the trapping of pollutants which leads to very poor air quality. Thus while the popular notion is

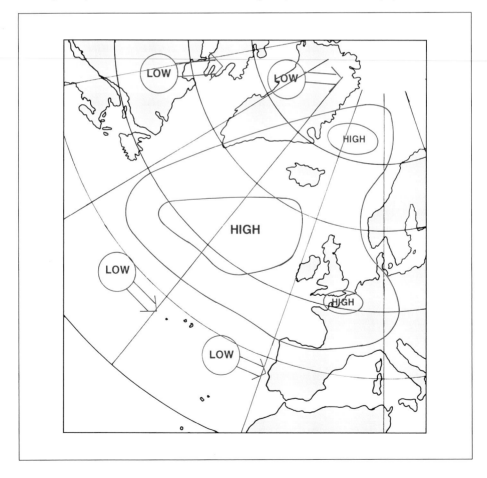

that anticyclones bring sunshine in many instances they do not.

February is the month which finds the longest blocks over Europe their average duration being longer than a month. This does not mean good weather for those under the winds blowing along their edges. It is the block which is an extension across Scandinavia of the Siberian high that produces some of Europe's worst winter weather. The intensely cold easterly winds are carried along the block's southern side entraining fronts and lows over southern Europe as they do so. The result is much snow and intense cold right through into Britain. The extremely bad winter of 1963, when people were walking across arms of the sea that had frozen solid, was such a one.

However what highs normally bring is light winds and together with clear skies they are the regions where the local winds are most likely to be found. Also they bring fogs especially in autumn as well as over-night frosts over much of the year excluding the summer months.

## Fog

The limits of visibility which are officially described as 'fog' depend on who the forecast is aimed at. For aircraft and in shipping and coastal waters forecasts there is fog if the visibility is below a kilometre. Otherwise visibility in sea area forecasts is described as follows:

*Good*    *More than 5 nautical*
          *miles (nmi)*
*Moderate*   *2 to 5nmi*

*Poor*    *1km to 2nmi*
*(For aviation and in coastal waters*
  *reports there is the category of Mist*
  *1 to 2km.)*

For land areas where vehicles are involved we have;

*Mist*       *200 to 1,100yds*
*Fog*        *Less than 200yds*
*Dense fog*  *Less than 50yds*

Fog that originates over land and that which forms over the sea are different. Land or radiation fog forms near the ground on clear nights due to the air near the surface falling to below its dew point temperature. When this happens the water vapour in the air has to condense onto minute dust and other particles and so forms a cloud on the ground. The depth of the fog layer may only be tens of feet but it tends to thicken with the night and is usually burned off by the sun in the morning. When it does not disperse in the morning a cloud blanket has crept in over-night and so prevented the sun from getting to the ground. Sunlight has to penetrate the radiation fog to clear it as such fog clears not because the sun heats its top but because up-currents occur from the ground.

It can be said that rain clears fog. When rain is forecast, even some hours ahead, you can be pretty sure that fog you now have will go before the rain arrives. If it does not then the falling rain will improve the visibility. It is not necessarily the weather regime ahead of rain which leads to the clearance but the wind that goes with it. Any significant increase in wind strength will clear fog but sometimes all that happens when the wind begins to blow is that the fog layer lifts off the ground and becomes low stratus. The only situations where fog may persist or intensify when the wind gets up are near the centre of lows at sea and over high ground.

The risk of fog depends on locality, time of day and time of year. This is shown best by a diagram like Fig 14.3 and indicates that at a low-lying place like Liverpool Airport during the summer months thick fog is confined to the hours surrounding sunrise, whereas for the rest of the year it can survive all day. It is worth noting when setting off for work on a potentially foggy morning that with a start at 07.00–08.00 it may not appear too bad but the fog thickens towards 09.00. This is because industrial and other polluters of the atmosphere are by then beginning to add their contribution to the particles that help to thicken the fog.

On the other hand because towns tend to be heat islands it may be that going via a town as an alternative to by-pass or country roads will make for a less worrying journey. Low fog forms readily over fields and drifts over roads at times when towns are quite clear. Added to which there is the miraculous way in which sodium street lighting seems to banish fog just as it banishes falling snow.

It is important to realise that while fog cannot survive in moderate to fresh winds it will also be loth to form in flat calm. However a slight wind increase will sometimes lead to a sudden production of widespread fog. In other words fogs like some, slight, wind.

Sometimes fog from elsewhere will drift over a previously clear locality. This is called advection fog – where advection means moving sideways as opposed to convection which is upwards.

When there is general low cloud we can expect hill fog and must allow for it when travelling into the hills or mountains. Sometimes a moist airstream will be reduced to its dew point by having to blow up hill slopes and so forms upslope fog. However on the leeward side of a ridge we can usually expect fog, formed on the windward side, to clear.

Sea fog forms when moist air is cooled by having to flow over colder water. Thus maritime Tropical (mT) air is most prone to develop sea fog which can be widespread when the air is very moist. If not so moist we can expect fog banks or patches to form which are a hazard because they can, by advection, suddenly envelope a craft at sea. At other times only the lifting of the air over a coastline produces fog although such air could well go on to produce upslope fog further inland. Often, when the conditions hover between fog and no fog, the first signs of sea fog appears over the beaches and the fog then develops seaward.

Another hazard of coastal fog, especially in spring and early summer, is the ability of developing sea-breezes to

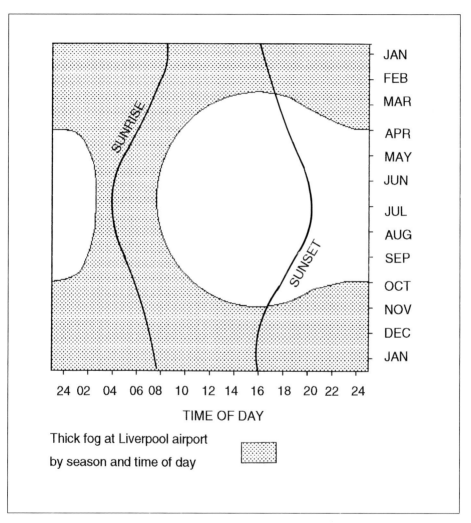

JAN
FEB
MAR
APR
MAY
JUN
JUL
AUG
SEP
OCT
NOV
DEC
JAN

SUNRISE

SUNSET

24 02 04 06 08 10 12 14 16 18 20 22 24

TIME OF DAY

Thick fog at Liverpool airport
by season and time of day

bring in an offshore fog bank. One moment the beaches are bathed in sunshine and the next are blanketed in cold clammy fog. This is enough to ruin any trip to the seaside but will not be helped by sitting in a traffic jam as everyone decides at once that it is time to

*Fig. 14.3 This diagram sums up the normal seasons and times to expect fog.*

pack up and head home. However, if you do, you will often find sunshine and warmth in the hinterland because the fog is only going to survive for a short distance before being burned off.

Not all sea fogs come on mT air. When a NE wind blows from Scandinavia across the North Sea there is a sufficiently long sea track for the originally dry air to acquire enough moisture to form coastal fog on the British east coast. Such fogs can be dense and have acquired the name of 'haar' probably because they get as thick as your hair.

Then there are regions of almost permanent fogs such as the Grand Banks off Newfoundland. Here the fogs are due to the clash of the cold Labrador Current with the warm Gulf Stream. Anywhere there are cold upwellings in the ocean will be a potential area for the production of sea fog.

### Frost

Because air is such a poor conductor of heat and only really exchanges heat with the ground so it may often be freezing on the ground but not in the air. When the over-night forecast chart shows minimum temperatures around a degree Celsius or so, they are showing you air temperatures. Thus it will probably be freezing on the ground. Under these circumstances any slight wind will help to keep plants frost-free but those in little hollows or pockets sheltered from the wind will show patches of *hoar frost* by morning.

A forecast air temperature of zero or any minus figure will mean that the ground will be several degrees colder. It is then that there may be icy patches on the roads. The most potent producer of black ice on the roads is the winter cold front which clears during the evening leaving the roads etc wet. The clear skies and polar air behind the front do the rest. Luckily only on the rarest of occasions do we experience glazed frost. For this we have to have rain falling onto chilled surfaces. Everything, leaves, branches, pavements, roads etc become encased in a skin of clear ice and any form of movement becomes dangerous. Air temperatures can be as high as 5°C and glazed ice can still occur on a road surface providing the latter is around −4°C. However only the sudden arrival of warm air in the morning of a radiation night could produce such a scenario.

The air temperature has to be many degrees below freezing and continue so for many days before the sea will freeze. Any impure water will have to be cooled below the normal freezing point before it will form ice. Salt sea water will only normally freeze along the shores when it floods in the morning after a radiation night. In most of the temperate latitudes it takes a very exceptional winter before the creeks and inlets freeze over sufficiently to support the adventurous.

In such deep winters anyone who has to go to sea should make sure they have chipping hammers on board to remove any *rime icing* that occurs on stanchions, rails etc which will, if not removed, severely increase the top-hamper.

Freezing fog is a particular hazard as it forms rime icing on all cold metal objects. For freezing fog to form we need

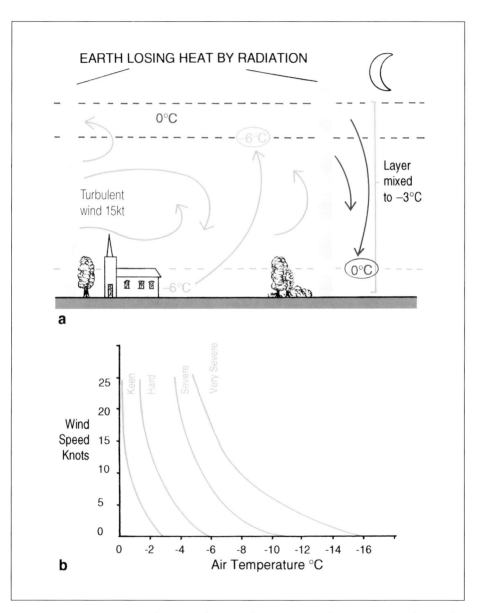

Fig 14.4 (a) How a wind mixes a layer of air and leads to a hard to severe frost. (b) How to make up your mind as to the degree of frost expected.

an exceptional amount of moisture in the air on a radiation night. When there is normal humidity the fixing of moisture in hoar frost robs the air of water vapour and often does not leave enough to form fog. We used to have a saying when I was a forecaster 'If we frost we don't fog and if we fog we don't frost'. The first has been explained and the second means that if the cooling of the air leads to fog forming then the latter forms a blanket and prevents frost.

The air temperature is not the only criterion as to the degree of cold which really means the ability of a cold night to freeze the plumbing. When there is wind that is below freezing, frost becomes more severe. This occurs because of turbulent wind mixing the cold air layers on the ground with those of an inversion higher up. In Fig 14.4a, if there is a moderate wind only the first thousand feet or so of air is mixed. The air near the surface is assumed cooled to –6°C while the inversion is at freezing point. Mixing of parcels of air from these two and intermediate layers will bring the whole layer to the intermediate temperature of –3°C. That this will lead to a hard to severe frost can be gauged from Fig 14.4b. It is not all that much of a frost when the air temperature is –3°C and it is calm but let that temperature combine with a 15 knot wind and the frost becomes a hard one equivalent to about –7°C with no wind.

Frost in autumn soon melts in the morning because the summer heat has penetrated to deep levels and is now working its way back to the surface.

Likewise early snowfall will not lay around for long. The snow is not usually melted from above but from below obeying the law that heat flows from where it is warmer to where it is colder. It may often be noticed when hoar frost covers pavements and then melts that there are corridors where it lingers. On investigation you will find that these frosty swathes are where someone has recently dug up the pavement. The loosely packed earth in the trench is a poorer conductor of heat and so the surface is insulated from the deep store.

As the winter progresses, so the stored heat becomes progressively lost until in February the ground is now deeply cold. Snow in February is often here to stay and frost penetrates well into the ground. Now there is less heat coming to the surface and snow and ice have to be melted either by the sun or, better, by the arrival of warmer air. Fast thaws can only be achieved by air with a tropical origin i.e. by the passage of a warm front and then the hazards are fog and floods.

Snow in spring is not all that uncommon but the sun has by now sufficient power to melt the snow from above and frosty mornings soon disappear for the same reason. It may seem odd to include facts about snow in a chapter on high weather but some of the deepest local snowfalls occur when little lows glide down the sides of established anticyclones. They are often sluggish and slow moving. They may even move retrograde i.e. from east to west rather than the usual west to east.

Chapter Fifteen
# Thunder and Lightning

Thunderstorms produce a primitive fear in people and no wonder when lightning has the power to kill and maim and there's nowhere to hide.

I was reading a widely distributed booklet on weather the other day which said without qualification. 'There may be lightning without thunder'. As lightning without thunder is impossible, what I hope they meant was that lightning may be seen but you may be too far away to hear the thunder.

Thunder is the result of the expansion of the air along the path of the flash. The super-heated air explodes outwards and sends pressure waves reverberating through the clouds. This is why the noisiest thunder may not come from the most dangerous storms. Storms can break out well above Earth's surface and then the lightning will often be from cloud to cloud, or strung out along the cloudbase. Such storms produce masses of rolling thunder as well as what is called 'sheet' lightning. Sheet lightning has long been recognised as not dangerous and this is understandable when we realise that there is no such thing as sheet lightning. When forked lightning flashes from cloud to cloud sometimes only the reflection can be seen bounced off chasms in the clouds and it looks like a sheet. However that flash had to be miles away for you to see it and the flash was not to the ground anyway. Thus it was perfectly harmless.

Not so the cloud-ground flash that nearly gets you. Then there is no rolling thunder clap – just one big bang and probably a sizzle as the lightning leaves the ground.

Those most prone to being struck are isolated people on golf courses and similar. Likewise it is isolated trees that are most often struck but they need to be broad-leaf trees and not conifers.

The suggested reason why conifers are rarely struck is that they have millions of needles and so millions of points. It is a fact that a pointed object cannot retain electric charge. What is called an 'electric wind' of ions (charged particles) streams from the point and the ions are of opposite sign to the charge in the thundercloud above. The electric wind created streams up into the base of the cloud and nullifies some of the charge there thus often lowering the potential for a flash.

This is the theory behind lightning conductors. A pointed object connected by a thick copper band to a good earth will create say a positive electric wind and may thereby serve to prevent a flash. However if a flash does occur the electric wind's path will make a channel to guide the flash to the conductor which then passes harmlessly to earth.

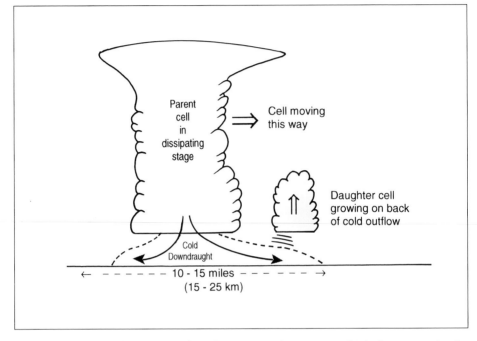

*Fig 15.1 How a dissipating thunder-
storm cell breeds a daughter cell nearby.*

A yacht with a similar conductor
above the masthead and connected by
the straightest path possible to an earth-
plate in contact with the sea will create a
cone of protection around it which is
much wider than the yacht. Whether so
protected or not, yachts are rarely struck
which seems odd seeing that they are
isolated on the sea. However that does
not stop them occasionally being struck
just as it does not stop the occasional car
being struck. However without doubt
one of the safest places to be in a storm
is the car. For one thing being
surrounded by metal creates a form of

'Faraday's Cage' which the great scientist
and father of electricity used to use to
protect himself from the more lethal of
his experiments. It makes a good hide-
out for worried pets too.

One of the reasons why yachts at sea
are rarely struck by lightning is that
thunderstorms are created over land and
only by chance do they drift over the sea.
However we must point out the different
kinds of storm there can be.

**Thunderstorm cells**
The thing we call a thunderstorm and
which erupts on hot sultry afternoons is
in fact a mass of individual storm cells
going through a life-cycle from birth to
maturity. The storm would soon die out if

the mature cells did not grow 'daughter' cells next to them. They do this when their cold down-draughts spread out around their bases and lift the warm, humid air into instability. (Fig 15.1) The daughters take over the role of the parents and soon produce strong updraughts that lead to very heavy rain and hail. It is the cold air dragged down by the deluge which spreads out and induces a sudden increase in the wind to perhaps 30–40 knots from the storm. Previously the wind was maybe 5–10 knots towards the storm. This zephyr that blows sluggishly towards the coming bad storm is a wind of the storm's own making, being air drawn in to feed the updraught. Thus the old saying that 'the storm comes up against the wind' is true but only in respect of the worst storms.

Contrary to popular belief a thunder cell is past its prime when it grows an anvil. (Fig 15.2). Cells that are producing hail, lightning and thunder have rounded cauliflower-like tops. However anvil-headed clouds are still the trademark of the coming storm because they are the dying parents of active daughters growing between them. (photo 15.1)

The speed of movement of a thunderstorm of this kind is usually about 20 knots but you cannot find that out from the surface winds which will be very variable with some prodigious gusts and wind shifts that defy any reasoned explanation. The only signs you have of the way the storms will move is from

*Fig. 15.2  Three stages in the life-cycle of a thunderstorm cell.*

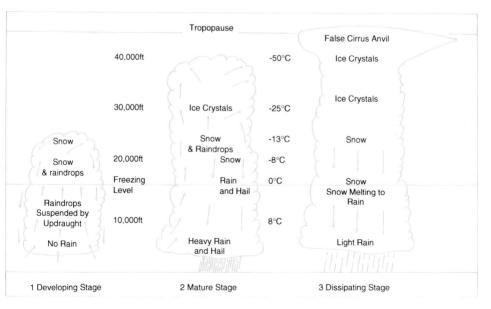

| | | 1 Developing Stage | 2 Mature Stage | 3 Dissipating Stage |

*Photo 15.1 A thunder cloud is at its most virulent before it develops a true anvil. This one with the cauliflower top is producing heavy hail.*

observing the flight of alto clouds before they arrive.

### Timing the storm

As light travels a hundred thousand times faster than sound so the sound of thunder arrives well after the sight of the flash that caused it. The rule to find how far the flash was away is:

> *Count the seconds between flash and thunder and divide by 5 = distance in miles (divide by 3 for km).*

We hear people say 'Oh the storm has moved around and come back' when a second set of cells moves in after a primary one has moved on. This is a mis-apprehension caused by the idea that a storm is a single entity. Once you under-stand the cell theory and realise that storm cells are continually erupting as the storms move you see that storms do not 'move around'. The second outbreak is just that – a new set of cells following the former ones. Also a storm may arrive over you rather quicker than you might think from the above timing rule because there may be new daughter cells breaking out between where you saw the flash and yourself.

Sometimes at night you may see the flashes but not hear the accompanying thunder. This is because thunder rarely travels further than ten miles. It will travel further across wide valleys from storms on opposite slopes but on lowlands or over the sea about ten miles is the limit.

**High level storms**

Another form of storm which affects northern France, the Low Countries and southern England is one which breeds over France in the heat of the day and moves up across the English Channel and surrounding areas. Often the first signs of this on a sultry morning are easterly winds, poor visibility amounting almost to fog and special forms of alto-cumulus (photo 15.2) called *floccus* and *castellanus*. These tend to form in lines along the wind at their level and I have

proved by experience that the way these clouds move with respect to the surface wind is a good guide to whether there will be thunder to follow or not. Sometimes we see floccus and castellanus and no thunder follows, but if the wind at alto cloud level is crossed to the surface wind for deterioration we can expect thunder to break out later, but if they stay

*Photo 15.2 The kind of sky to expect before thunder. Note the turret-tops of castellanus in the centre of the picture and the small round and yet chaotic globules of floccus at top left.*

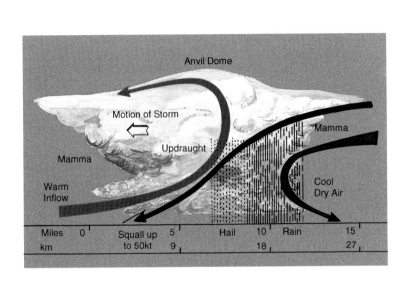

*Fig 15.3  Anatomy of a supercell.*

doggedly in the same direction either there will be no thunder or the storms will be delayed into the evening or night.

## Supercells

There is however a storm of the very worst kind that is one vast single cell. It is called a 'supercell' and it is supercells that produce giant hail-stones. There was one, which I did not experience but which I remember very well. I remember it because my father saved the hail in the freezer until I could get to see it. It was not round as hail is supposed to be. It was great nobbley lumps of ice which broke panes in the greenhouse. Such hail does occur occasionally and it then causes considerable risk to anyone who is not under adequate protection. In one case in the south of France dinghy sailors who tried to shelter under their sails only

found that the stones went straight through the material.

No one is quite sure how the supercell originates but the squall that comes out of its leading edge may be 50 knots or more. (Fig 15.3) Here we have one vast anvil dome and the storm seems to feed off the surrounding air which needs to be very humid.

*Thus storms can be divided into categories*
1) *Airmass storms which grow individually in their own personal spaces.*
2) *Frontal storms where the bases of the thunderheads may be many thousands of feet off the ground.*
3) *Heat storms which break out in thermal lows in summer.*
4) *Supercells which occur maybe once in ten years and then often not in the same locality. (More often in the U.S.)*

Chapter Sixteen

# Some Fundamental Principles

There are certain principles which will help anyone to make more sense of the weather and the way it acts. First of all we will put down some rules which will aid in sorting out the situation and follow them with some more basic physical ideas.

### Fundamental Principle 1 (FP1)
### Establish the kind of airstream

From the characteristics of airmasses given in Chapter 4 it should be possible to make up ones mind as to what kind of air is affecting you now and will do so in the immediate future. Remember that an airmass will not change until a front passes but that passage over land or sea can modify the airstream near the ground very considerably. When looking at a weather map you need to look back down the isobars to where the air is coming from.

To illustrate this look back at Fig 6.1. For Britain and most of the Atlantic coast of Europe the airstream is coming from the sea areas off Iberia but it is not maritime Tropical air. We see this because the air is ahead of a warm front and a warm occlusion. Therefore this will be a modified mP airstream and we would expect fair visibility and some cloudiness which was likely to be in the form of heap cloud or maybe strato-cumulus.

Over the central region of Norway/Sweden we see how we can get two different airmasses next to one another without a front interposing itself. Over southern Scandinavia the air is the same as just described for Atlantic Europe but over the Gulfs of Bothnia and Finland the air is coming round the complex low from cold northern seas. There is mainly anticyclonic curvature over Denmark and surrounding waters while over the Baltic States it is cyclonic. We could expect snow over the Norwegian/Swedish backbone of mountains with a much drier regime over the parts of Scandinavia that lie in its lee.

Over the eastern seaboard of the United States the air is coming from the cold seas around Greenland and so will be cool and moist and therefore there should be good visibility but much heap cloud. Skies could well be clear over Labrador under the influence of the anti-cyclone there. Over northern Labrador the air will be dry and under anticyclonic conditions might be expected to be largely cloudless but on glancing north-wards we see that this is in effect 'warm' air and so maybe we can expect extensive sheets of stratocumulus.

These few examples illustrate the technique of looking back to where the air is coming to establish its likely characteristics.

## FP 2 Consider possible changes

Take a limited period ahead – say six hours. What can happen in that time? If it is breakfast-time then six hours brings us into the early afternoon so we have to add the diurnal changes to the weather we have now. If there is any chance that there will be sunshine during the morning we must expect both cloud and wind to increase. If there are weather systems in the offing what will they do to the wind direction and speed? Is a front likely to pass in that time? If so, is it warm, cold or an occlusion. What will that do to the weather later?

If it is a cold front or occlusion, will there be a sharp drop in temperature overnight. Is it possible that there will be icy patches on the roads by tomorrow morning? If it is a warm front what chances of fog? Here are a few questions to ask yourself depending on the situation and the time of year.

a) *Will it rain (or snow in winter)?*

b) *How hard is it likely to blow. If contemplating going offshore sailing how hard is it likely to blow outside the confines of the land?*

c) *Will there be fog or will low cloud shroud the hills?*

d) *Are thunderstorms likely?*

e) *Will there be frost?*

f) *If the situation has been set on one particular kind of weather for some days is it likely to change today? For example in a run of days of snow-cover what prospect of a thaw today?*

You may not be able to answer these questions but sometimes you can and if not they should prompt you to seek a forecast and see what the pundits say.

## FP 3 The wind veers with height when a warmer airmass is on the way and backs when a colder one is coming

The ramifications of this fact have been covered in Chapter 5 and we need say little more here. However do not neglect the fact that this is a very important and useful rule for the person who would foretell the weather.

## FP 4 Winds blow anti-clockwise about lows and clockwise about highs (reverse this for the Southern Hemisphere)

What this rule does is set you correctly in the local pressure pattern. From the wind direction you have (and modified to allow for the way the wind blows at an angle across the isobars) you can stand back to the wind and find where a local low pressure centre is situated.

## FP 5 Weather worsens under cyclonic and improves under anticyclonic curvature of the isobars

It is a physical fact that air which blows round bends has either to rise (cyclonic) or sink (anticyclonic) and as the former means more cloud, rain, showers etc so the trend is towards worse weather when such a set of isobars falls over an area of interest. Equally the opposite curvature leads to air sinking from high up. This may not break a low cloud layer but can lead towards it when upper clouds are eroded away. The sun (when high enough) can then 'burn-off' the low cloud layer.

## FP 6 Short forecast – soon past. Long foretold – long hold

In human relations the acquaintance that builds slowly often endures while the person who was all over you at first passes in the night. The same goes for weather. The warm front that takes its time coming will usually herald bad weather for an equally long period while the frontal cloud that builds in a few hours will pass just as quickly. The trough which arrives almost out of the blue will just as quickly be gone. The sunshiney shower won't last half an hour. It is a universal law of nature.

## FP 7 Intensity times time tends to be constant

Sometimes it starts to suddenly deluge with rain. Such a deluge cannot be sustained if for no other reason than the clouds will use up their available moisture. So the more intense the rain, snow or hail the shorter time it is likely to last. The slow gentle rains of warm fronts and occlusions last for many hours while the sharper cold frontal rains do not last so long.

There are many provisos to such a blanket rule – it is only true in general. For example, the above refers to lowlands. On the upslopes of hills and mountains the deluge may go on for a very long time. Waves running along cold fronts can slow them up and lead to more extended precipitation. Yet on the whole the rule is right.

The wave lows just mentioned being small and with light winds will run past in a very short time whereas the major low, which throws its influence over thousands of square miles will be with you maybe for days.

## FP 8 Lows that deepen slow down and vice versa

This is really the conservation of energy principles which says, in this case, that if a low is using some of its available energy to deepen it cannot have as much left over to travel. There was a good example of this in August 1979. (Fig 16.1). A low left Nova Scotia on Saturday 11th with a central pressure of 1004mb and made an average 45 knots to mid Atlantic without change in central pressure. It then began to deepen so that it was 995mb when 600 miles off south-west Ireland at midday on Monday 13th. During this deepening it slowed to 33 knots. Then in the next twelve hours it deepened further to 982mb during which time it travelled at a slower 30 knots. So we see as it deepened it slowed up.

This was the low which was responsible for the so-called Fastnet yacht disaster when fifteen yachtsmen were drowned taking part in the Fastnet Yacht Race. As well as this loss of life five yachts were also lost, twenty-four were abandoned and 136 crew members had to be rescued.

Once ashore over Ireland the low stopped deepening and moved into Scotland at an enhanced 40 knots.

Some lows may be so virile that they will deepen and travel fast at the same time but on the whole the above rule is obeyed.

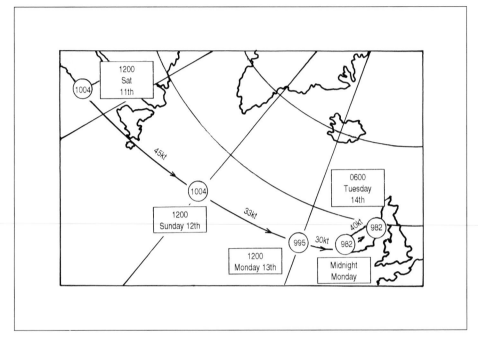

*Fig 16.1 The low that devastated the Fastnet yacht race in 1979 followed the track shown here.*

### FP 9 Strong weather stems from strong temperature contrast

At the moment as I write this, it is blowing very strongly outside but nothing like as strongly as it is over Scotland and northern England where the forecast is promising winds up to 100 mph and rain and blizzards as well. It is mild down here in the south of England, 11°C in fact very mild for mid-January. The weather is due to an unprecedentedly low depression off Iceland whose central pressure of 920mb may be an all-time record. It is very cold in the north-

ern parts of this low and we see a very big temperature contrast between the warm and cold air. It is this contrast on which the low depends for energy to put into its winds and weather.

Incidentally there have, in recent years, been depressions whose central pressures have fallen way below the magic 950mb which is supposed to be a very deep low. The depression which moved up from Biscay across England in October 1987 and produced the 'hurricane' over northern Spain, northern France and south-east England fell to 950mb – the lowest pressure ever recorded over mainland Britain. We may here be seeing the tangible results of the promised 'global warming'. The tropics are getting warmer while the polar regions will not change their temperature much. The result is stronger temperature contrast between mT and mP airmasses and so very intense depressions.

### FP 10 What goes up must come down

Nature abhors a vacuum and so when we say that depressions are regions where air is mainly ascending we have to remember that such a thing could not happen unless air came in below to replace that which has gone up. It is the same with anticyclones. Air cannot go on sinking from way aloft unless it is replaced up there.

Some loss or gain of air is allowed by the pressure changes but depressions have to be fed with air from the surrounding anticyclones or we would indeed have a vacuum down here and find it difficult to breathe. Likewise an anticyclone would soon stop feeding air down to us if it did not get a replacement.

On a smaller scale the exceptional visibility which has been recognised for centuries as presaging bad weather comes from clear air sinking ahead of and under warm fronts to help compensate for the ascending air along the frontal surfaces themselves.

On an even smaller scale when thermals rise from warmed surfaces there has to be sinking air all round them to replace the air lifted from the surface. This up-and-down motion is a very important component of the variations we find in the wind when heap clouds

ride the sky.

### FP 11 Weather moves in a continuous manner

We see an example of this in Fig 16.1. Despite the fact that all weather systems are made of gases and vapours with no solid substance to them we move them about as if they were rigid bodies. If you think about it, much of the air which was circulating round the low over Nova Scotia was not the same air as was eventually blasting the Fastnet fleet. If you had labelled an air molecule Fred over Nova Scotia would Fred have been still found in the low when it got over Scotland? It is doubtful. Weather systems are the propagation of a condition and the air and water vapour that feed that condition are constantly on the change. Yet the idea of continuity is the one on which any useful weather forecasting rests. We can move lows and highs around – give them letters to identify them and generally expect them to obey certain rules. Fronts are the same. Fronts will have air and moisture moving through them all the time but we can move a front across the weather map as if it were a more-or-less rigid wire. It is the condition that constitutes a front that we move regardless of the fact that the stuff of which the front is made is constantly on the change.

### FP 12 Heat seeks cold

Here we have a scientific principle of the highest importance in understanding weather processes. An obvious example

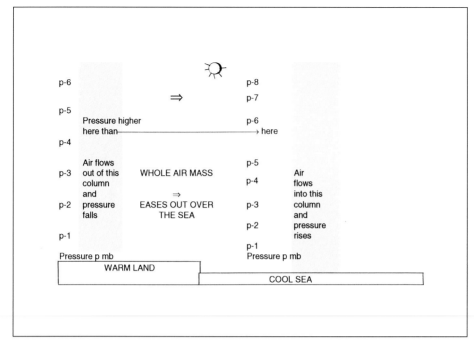

*Fig 16.2 The real way that sea-breeze starts is for the air over the land to expand upwards taking the pressure levels with it. Thus if the surface pressure over the land and sea is P mb at first so the warm column has higher pressure at any level than the cold one. This induces the air to move out of the land column (divergence) and into the sea column (covergence). The result is a pressure difference that makes a breeze blow from sea to land.*

is thermal currents where parcels of air warmed above their surroundings lift off to seek colder realms aloft. Any lump of air that is warmer than adjacent air will try to move towards the colder air. If, as they rise and cool, they continue to find colder air above them they will go on rising and (with the help of latent heat energy released within clouds) eventually only stop when they reach the tropopause. Then we have Cb clouds with anvils. Most heap clouds do not get that big because they find an inversion layer warmer than themselves. Now the thermal air is cooler and the inversion layer's air tends to sink in accordance with the above principle. This is where Cu clouds stop rising.

In the same way, the clouds along warm frontal surfaces will be layered because they are always trying to sink through the frontal surface as we have warm air lying over cold. A front is a big sloping inversion.

Sea-breezes and land-breezes seem to go against the principle but that is only because they are the result of a more fundamental air motion which does obey it. Sea-breezes develop only because pressure is lowered over warm land compared to the adjacent sea. What happens is that as the sun warms the land so the air layers there expand upwards. The pressure levels rise with them so that as in Fig 16.2 there is higher pressure at say 20,000ft over the land than over the adjacent sea which has not changed. Air eases out over the sea under this pressure difference so relieving the weight of air over the land and increasing it over the sea. Now surface pressure is higher over the sea than over the land and a breeze blows to try to equalise the difference. So once again heat over the land seeks cold over the sea and the breeze is the result. Similar considerations apply to land breezes.

Finally, in these few examples, consider how the wind goes down with the evening. The earth radiates its heat and the air in contact gets colder than the air above. So the trend is for the air layers just above the surface to sink. This begins to kill the turbulence that has mixed up the air during the day and fewer chunks of higher speed air come down to the surface. So, as the process continues and the cold layer gets deeper, the surface wind is more-and-more cut off from the wind at higher level. As the surface wind has to blow against the friction of trees, buildings etc so, without help from above, it dies away, often to nothing.

## FP 13 Cold over warm is unstable. Warm over cold is stable

This is in some ways a re-statement of the above but is directly applicable to many weather situations. For example, as winds come down from polar regions

behind retreating depressions they carry mP air across sea temperatures that are all the time increasing. So cold air finds itself over relatively warm sea. The result is masses of big shower clouds which will invade the land when they reach it. The air becomes more and more unstable as it comes south.

Over land by day, especially in spring, mP air will be bright and clear in the morning but develop many showers as the day proceeds. This follows because the sun has great power to heat the land especially when it is as late as April or May but the spring air is still relatively cold.

There are other less obvious results of instability. Sometimes layers of dark alto-stratus or altocumulus cloud will suddenly produce showers. The reason can be traced to the sun which is shining on the top of the cloud layers and warming them up. This warming initiates Cb clouds growing out of the As or Ac layers and sometimes these high level heap clouds can become thundery.

The latter is something we find mainly on spring and summer days when the sun has great power. In autumn cool surface winds sink off the land onto coastal waters still warm from the summer. Again we have cold air over warm water and the result is coastal showers in the early hours which may persist after dawn and sometimes invade the land.

The stable case has been discussed in Chapter 8 because it is the inversion situation. In unstable conditions warm air freewheels to as high a level as it can but in the stable case all changes have to be forced against the air's natural inclination. So when mT air comes up across lower sea temperatures it tends to sink and clouds get down near the deck while fog is an ever-present possibility. Thus spring and early summer tend to be foggy periods in Atlantic Europe and surrounding sea areas because the tropical air that comes on the SW wind is warm and humid but the sea has not yet acquired its summer warmth. Go further south and the sea temperatures are that much higher so that fog is much less of a risk.

### FP 14 Wind equalises

To explain this principle think of a calm, cold night with ground frost. The air layers just above the surface will be very stable and there will be a marked change in temperature with height. It may be several degrees below freezing on the ground, around freezing in the air and maybe several degrees above freezing at say 500ft. That is what happens when the air is stabley stratified. There will be little if any surface wind.

Now the sun gets up and begins to warm the ground but not the air above it. This immediately begins to reverse the stable situation near the ground. An unstable trend begins to appear as the ground communicates its heat to the air close to it. Small eddies appear and begin to mix the lowest layers. The more the sun heats the ground the higher these eddies penetrate and soon the over-night inversion layer is being broken. This allows the sinking currents of the eddies to bring down stronger wind that has previously been locked away above the inversion and the surface wind begins to pick up. However at the same time this mixing will bring warmer air down to where it is coldest and lift cold air higher to where it is warmer. The result is that the whole layer that is mixed becomes much closer to an average temperature. The surface air warms up while the higher air layers cool down. Thus wind, by producing turbulence, tends to break down strong temperature contrasts in the air layers.

The same goes for wind that may get up during the night due to the incursion of a new pressure gradient. There may be frost in the first part of the night but the arrival of the new wind mixes the warmer air above with the cold surface air and what might have been a hard frost does not occur.

On very cold nights during which the wind continues to blow, the stirring up of the lowest air by the eddies in the wind stops the temperature from dropping as low as it would had it been calm. However the degree of cold is much greater as the wind can penetrate into roofs etc and so produce freezing of pipes and tanks.

Chapter Seventeen
# Conclusion

We live in an age when the computer is king. In particular the vast high power machines that churn out the weather forecast charts and many other things as well have produced a revolution in weather forecasting. It is now routine for a forecaster to appear on the TV on a Sunday and give you a very good account of what the weather will do for most of the rest of the week.

However there are many facets of the weather that the brave new forecasting world had difficulty with. It is not yet able to be absolutely accurate about when rain will appear over you and how much there will be. Fronts that were expected turn up later, or earlier than thought. There is still work for the private weather forecaster to do.

However this happy breed are now in a much stronger position than they used to be. By keeping a weather-eye on the TV forecasts where actual charts are shown one can know very accurately what is expected and so from your knowledge of weather systems find out for yourself whether the expected weather is running true to form or not.

Once you understand the processes of weather better, it is amazing how much more you can read into the hurried and almost frantic deliveries that the poor TV presenters have to make. You can use the 'video and pause' method I have outlined on page 146 to allow yourself time to really study the charts and make up your mind if the isobars are indeed cyclonically or anticyclonically curved over you and so draw some local inferences that they will not refer to on the forecasts.

We are in a period when weather records are being broken all the time. We referred in the last chapter to the record-breaking low pressure of the Atlantic low of 10 January 1993. Out at the weather ship off the west coast of Scotland the winds blew as high as the magic 100 knots and the waves were over 30 feet high. I have never seen such a system in all my sixty years of being interested in weather and it is happening all the time.

I turn up the July 1992 edition of the Royal Meteorological Society's magazine *Weather* and find a letter which starts 'The three months of winter 1991/1992 (December, January, February) have seen a persistence and intensity of anticyclonic conditions without precedent for at least 40 years in southern England.' The rains that have followed that drought period have been almost as prodigious as the length of the drought before it with much flooding and intense storms.

The *Daily Telegraph* of 10 December 1991 said 'The Greek army was put on alert yesterday as freak snow and rain storms continued to sweep the country, causing more than 15 deaths and disrupting traffic and communications in mountain towns and villages . . . Food and supplies will have to be dropped by air to more than 120 villages and small towns . . . Traffic crept to a standstill in Athens as abandoned cars and uprooted trees blocked highways and major city routes.'

Another report on 14 February 1992 said 'Eight people are dead and several missing after two days of torrential rain caused havoc in the Los Angeles area. Dozens of cars and mobile homes were swept out to sea when flood waters overwhelmed a caravan park at Ventura Beach . . . The storms have deposited up to 10 inches of rain since Sunday, but have brought little relief from the drought now in its sixth year, officials said. Southern California gets most of its water from reservoirs hundreds of miles away, and they have missed the brunt of the storms.'

We could go on with such a catalogue as one after another weather disasters occur or records are broken. The weather is a constant source of interest as well as danger to our well-ordered lives and yet the modern methods are making life under the elements safer. Hurricane Andrew may have been the first one of the 1992 season but that did not stop it being one of the most vigorous of all time. Winds were up to 160 knots and there was extensive damage to property but only some loss of life. For that we can thank the satellite, the computer and the man on the forecasting bench who can now give good warning to threatened people to evacuate. The modern fore-casting systems earn their money despite the conviction of many that they still get it more wrong than right.

In the Americas the winter period of 1997/98 was subject to the extreme weather events that go with the so-called El Niño effect. This is due to unusual sea temperatures that turn up every so often in the Pacific Ocean. There is firstly an increase off the coast of Peru around Christmas (hence El Niño, referring to the Christ child) and at the same time colder than normal seas in the North Pacific.

As a result there were, amongst many anomalous weather conditions, droughts in South America, devasting tornadoes in Florida and other Southern States as well as unusually snowless winters in the Eastern USA. Both America and Europe had remarkably warm Februaries. There is an old saying which goes 'All the months of the year curse a fair Februeer.' In Britain, April 1998 was the wettest on record with unprecedented floods in Central England. This may also be a knock-on effect of El Niño as well as reaction to an abnormality in the sunspot cycle.

Which just goes to show that no part of the world's weather is an island. One nation's drought is another nation's deluge and one of the great lessons meteorologists have learned in the latter years of the Twentieth Century is that the world's weather is linked – often in ways we do not yet understand.

So there is still room for the home-based weather forecaster and I wish you luck in your forecasting.

# PART TWO
# Information and Explanations

**Adiabatic** When 'dry' air, (air which is not saturated with water vapour) rises or sinks it does so adiabatically i.e. without gain or loss of heat. Rising dry air always cools at 3°C/1000ft of ascent. This is understandable if we think of a rising air parcel rather like a big balloon. It expands as it rises and as it cannot get heat from anywhere the heat it already has is spread out into a greater volume. Thus the temperature of the parcel falls. This rate is called the Dry Adiabatic Lapse Rate (DALR). When air sinks it conversely warms up at the same DALR even though now the temperature is doing the reverse of lapsing.

Thermals carry warm, moist air up to where the temperature falls to the dew point. At the latter temperature the invisible water vapour has to condense out as water droplets and a cloud forms. The process of condensation releases latent heat which works against the DALR and slows it down to a value about half what it was before. This new rate is called the Wet Adiabatic Lapse Rate (WALR).

**Anemometer** A device for measuring wind speed. Various kinds are to be found from simple plastic devices to precise instruments. The lay forecaster does not normally need anything too precise as for one thing the wind is always varying and what one wants is a value for the mean wind speed. Yacht chandlers are a good source of anemometers.

**Barometer** A device for measuring atmospheric pressure. The aneroid (literally non-liquid) barometer consists of an evacuated capsule which collapses slightly when atmospheric pressure increases and expands when it decreases. These small motions are amplified and transferred by linkages to a dial. The dial is likely to be calibrated in either millibars or inches or millimetres of mercury. The former is most useful as it accords with modern met practice.

## Conversions of pressure units

| millibars | inches of mercury |
|-----------|-------------------|
| 1050 | 31.00 |
| 1040 | 30.70 |
| 1030 | 30.41 |
| 1020 | 30.11 |
| 1013.3 | 29.92 = *Standard atmospheric pressure (760mm.Hg)* |
| 1010 | 29.82 |
| 1000 | 29.52 |
| 990 | 29.23 |
| 980 | 28.93 |
| 970 | 28.64 |
| 960 | 28.34 |
| 950 | 28.05 |
| 940 | 27.75 |

Check readings in millibars are sometimes given on weather forecasts when the barometer is steady for a while as it may be in anticyclones. Not much credence should be given to the legends of fair, change and rain although if the barometer is so low as to be in the stormy bracket then it very often is, and if it is high enough to be in the very dry category then the chances of rain are very slight. What actually matters is the *tendency* of the barometer (see below) and to give the pressure as it is now and what it has been in the past.

**Barometric corrections** Atmospheric pressure varies with temperature and altitude but for most purposes the former can be neglected while the latter drops by 1mb for each 10 metres (30ft) of ascent. Values on weather maps etc are for mean sea level and only if above 100m (300ft) is a correction worthwhile.

**Barometric tendency** (or pressure tendency) is, in standard met usage, the fall or rise in pressure over the past three hours. It is often tedious and nerve-racking to wait three hours between readings and so the following may be a help in predicting imminent wind increase (if you have not already got the wind)

1mb/hour means a likely Force 6

2mb/hour means a likely Force 7–8 gale

3mb/hour means a likely gale exceeding Force 8

These values are for those at sea. On land reduce the expectations by one or two Beaufort forces. However allow for gusts as strong or stronger than the above figures. After a low centre has passed closely over you and the barometer begins to rise sharply the above will also give a guide as to the strength of the new wind to come.

The following terms are used to describe pressure tendency. All refer to the change over the last three hours.

| | |
|---|---|
| Steady | less than 0.1mb |
| Rising slowly or falling slowly | 0.1 to 1.5mb |
| Rising or falling | 1.6 to 3.5mb |
| Rising quickly or falling quickly | 3.6 to 6.0mb |
| Rising or falling very rapidly | more than 6.0mb |
| Now falling, now rising | rising to falling or vice versa |

**Beaufort Wind Scale (See opposite page)**
**Celsius** is the scale of temperature that marks 0° when placed in a mixture of ice and water and 100° when in steam (at standard pressure). It is sometimes called loosely the Centigrade scale but technically this is wrong as any scale which has 100 degrees or grades between the ice and steam points is a 'centigrade' scale. Page 143 shows a conversion scale between Fahrenheit and Celsius.

## Beaufort Wind Scale

| Beaufort No | Description | Wind Speed (knots) | Land Signs | Sea State |
|---|---|---|---|---|
| 0 | Calm | Less than 1 | Smoke rises vertically Leaves still. | Mirror smooth. |
| 1 | Light air | 1–3 | Smoke drifts. | Scaly ripples. |
| 2 | Light breeze | 4–6 | Leaves rustle. Flags not extended. | Small wavelets. Crests do not break. |
| 3 | Gentle breeze | 7–10 | Light flags extended. | Large wavelets. Crests may break. |
| 4 | Moderate breeze | 11–16 | All flags extended. | Small waves. Some white horses. |
| 5 | Fresh breeze | 17–21 | Trees in motion. | Moderate waves. Many white horses. |
| 6 | Strong breeze | 22–27 | Tree branches in motion. | Large waves with foam crests. |
| 7 | Near gale (moderate gale: US) | 28–33 | Walking into wind difficult. | White foam from breaking waves blown in streaks. |
| 8 | Gale (Fresh gale: US) | 34–40 | Twigs break from trees. | High long waves. Spindrift. |
| 9 | Strong gale | 41–47 | Roof and fence damage. | High waves. Dense streaks of foam. |
| 10 | Storm | 48–55 | Trees uprooted. Structural damage. | Sea white. Violent waves. |

| –40° | –35° | –30° | –25° | –20° | –15° | –10° | –5° | 0° | C |
|---|---|---|---|---|---|---|---|---|---|
| –40° | –31° | –22° | –13° | –4° | +5° | +14° | +23° | +32 | F |

| 0° | 5° | 10° | 15° | 20° | 25° | 30° | 35° | 40° | C |
|---|---|---|---|---|---|---|---|---|---|
| 32° | 41° | 50° | 59° | 68° | 77° | 86° | 95° | 104° | F |

For those unused to Celsius the following is roughly how it feels.

All minus values to

| | |
|---|---|
| 0°C | Freezing |
| 5°C | Cold |
| 10°C | Chilly |
| 15°C | Mild |
| 20°C | Warm |
| 25°C | Hot |
| 30°C | Very hot |

**Cloud cover** is technically measured in octas or eighths of the sky covered but a simplified form is often used in forecasts where the sky is divided into quarters. Not all regions describe the sky the same way but here is a guide.

Less than a quarter covered
      = sunny by day, clear by night.
A quarter to three quarters
      = fair so long as there is no
      precipitation or fog.
More than three-quarters
      = cloudy.
A solid, dull, uniform layer
      = overcast.

Cloudiness damps down the diurnal variation in temperature and wind speed. Thus clouds act like a blanket. Its arrival on a frosty morning usually leads to the frost going very rapidly because the clouds are warmer than the surface. This normally signifies a coming warm front or occlusion.

Coastlines and hill ridges are often lines of division between cloud on one side and none or very little on the other.

**Crepuscular rays** are the rays which break through chinks in the cloud cover, often near sunset, and often look like searchlights. They become prominent because the light, reflected off cloud, beams down through a hazy atmosphere and so becomes more visible. They have no real prognostic value except to show that the cloud is broken and that the situation is at present stable.

**Deepening** refers to a low whose central pressure is falling with time. Lows may deepen slowly or rapidly and sometimes even 'explosively'. Explosive deepening occurs in some Atlantic depressions when the central pressure drops at a phenomenal rate and the winds become storm force or greater. The centre of such depressions may be a thousand miles away but they will freshen the winds over most of that area.

**Dewpoint** is the temperature to which a parcel of air must be cooled before it has to condense some of its water vapour into droplet form. This condensation has to occur onto so-called 'condensation nuclei' which may be motes of dust, chemical particles, ice crystals etc.

On the ground, when the temperature falls with the onset of night, the invisible moisture in the air will use the minute hairs etc on the surfaces of plants as condensation nuclei. The result is dew.

The dew point temperature is higher if the air is more moist. Then, because a parcel of air can only hold a certain quantity of water vapour at a given temperature, when the temperature falls the air becomes over-saturated with water vapour. It then has to condense some as dew. The term is used off the ground when referring to the temperature at which rising thermals cool sufficiently to have to condense their water vapour into cloud. The dew point is then the temperature at the cloud base.

Finding dew point is not truly feasible for the average lay forecaster involving a wet-and-dry bulb thermometer and a set of hygrometric tables.

**Evaporation** is the opposite of condensation and depends only on the temperature of the water. This is because the water molecules are in constant motion with a wide range of speeds. The higher the temperature the greater the speeds become. As only the fastest molecules can escape, which means those with most heat energy, so evaporation cools water quite rapidly. It cools volatile liquids like petrol much faster. You can prove this by sticking the end of a finger into petrol and feeling how cold the finger becomes as it struggles to supply the heat being taken away by the rapidly evaporating molecules. This has an important bearing on 'exposure' as the wet body will lose heat much more rapidly than when dry.

When we talk about say, a puddle 'evaporating' what we are referring to is the 'net evaporation' (E) i.e. the excess of evaporation (B) over condensation (C). (Fig. E.1). The latter depends on how saturated with water vapour the air is over the puddle.

So-called 'dry' air is not really dry as it contains water vapour but it is dry

*Fig E1 Explains about net evaporation E which is the difference between what evaporates B and what condenses C.*

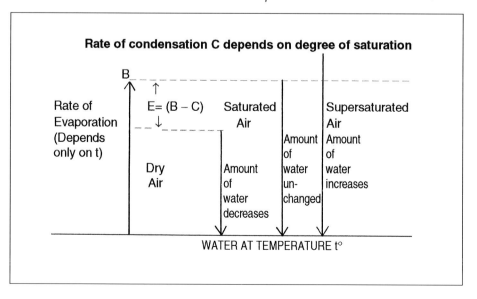

compared to 'saturated' air. When the air is saturated the rates of condensation and evaporation from water surfaces are the same so the water neither increases nor decreases. We have saturated air in fog.

The natural state of air in rain clouds is to be super-saturated. Now the rate of condensation greatly exceeds the rate of evaporation and ice crystals and water drops grow until they are too heavy to be supported by any updraughts and fall out of the clouds.

As wind tends to dry air by mixing wet air over a puddle's surface with dryer air above (see Mixing) so the net evaporation increases also with wind speed. Obviously both temperature and wind are used to clear car windscreens etc and washing dries best on warm windy days.

**Forecasts** can be had by various means which include the following:

*Domestic Radio*

One of the most accessible of the forecast media as you do not have to look at it like you do with TV and it is always with you in the car and the boat etc. National radio forecasts will give wide coverage with consequent reduction in detail for any given region. They will also give gale warnings and carry shipping and inshore waters forecasts.

Local radio will give a more detailed coverage for the local area and in yachting areas will pay particular attention to coastal winds and weather. In mountainous areas they will keep an eye on those aspects of the weather which affect climbers, walkers etc.

In the USA a 24-hour continuous set of weathercasts is broadcast by NOAA Weather Radio on one of the bands 162.40,162.475 or 162.66MHz

*Television*

This is the only readily available means of seeing actual and forecast charts. However as they are flashed before your eyes in double-quick time, so the only way the serious weather-watcher can study them is by what I call the 'pause and draw' method. You video the weathercast and then use the 'pause' facility to freeze the charts you want to view. That may be enough but if you want to really look at the situation I have found that by taping a large sheet of tracing paper over the screen you can trace off the chart. It need only be rough away from where you are, with more care taken in the local areas of interest. Once you have done this you will have the latest thinking of the top met brains both for now and for the future.

*Telephone*

Really detailed forecasts can be had over the phone both for land areas and for coastal sea areas.

*Facsimile*

The airwaves are crammed with charts etc being broadcast via radio weather facsimile. You need a good VHF radio with side-band facility, a decoder and a printer. Or you can hook up your PC to the radio and a decoder and print the charts through your computer.

The greatest problem the lay weather person will have is what to accept and what to reject of all the vast amount of meteorological verbiage they will be able to receive. The person with fax goes from having too little information to having too much. The charts which are Figs 5.2 and 6.1 were originally received via weather fax and as well as such actual charts there are forecasts of what the computer thinks the weather maps will look like six days ahead plus many more believable forecasts (called prognoses) for 24, 36, 72 hours ahead. There are plots of sea temperatures across the Atlantic and in the USA charts of the Gulf Stream and the Pacific Ocean. These and many other important aids to weather forecasting are available via a facsimile receiving system. A radio weather fax is almost a must for sailors contemplating going sufficiently 'foreign' to be beyond the range of forecasts in a language they can understand.

Telephone fax is now the means by which you can obtain information for land and sea, both actual and forecast. In Britain the Met Office's MetFAX service provides up-to-the-minute charts, both actual and forecast. There are, in addition, radar plots of precipitation, recent satellite images, special climbing and hill-walking forecasts, low-level aviation and inshore and offshore marine forecasts. The index page is on 0336-400-400. Other countries provide similar services.

Telephone forecasts are a useful alternative and in Britain the Met Office's Weathercall service covers the country in 26 areas. In France an automatic service, ALLO METEO FRANCE, provides a similar service for the Départements of France. In the United States the telephone recorded forecasts are to be found under Weather in the telephone directory and cover local areas including parts of a State but neither the whole State nor the whole country.

Today, with so many home computers connected to the Internet another way of obtaining weather information is via the appropriate web-sites. In Britain phone 01344 854153 for details of the service called MetWEB.

Navtex is a system which receives a form of fax signal but only prints out plain language shipping forecasts, gale warnings, navigation warnings etc. The small receiver and printer combined print out automatically and the system is far cheaper than facsimile.

Most countries will issue booklets which give the relevant details of their weather bulletins for yachtsmen, fishermen etc.

**Gales**
Gale warnings are issued with commercial shipping in mind and are for Force 8 (34kts mean speed) or more, or when gusts are expected to reach 43kts or over, even though the mean wind speed does not reach 34kts.

Strong wind warnings for coastal small craft are given over local radio and are for Force 6 to 7 (over 22kts) and in Britain cover a five-mile wide strip out from the coastline. Even Force 6 is too strong for many small craft and conditions where the wind may exceed Force 4 (over 15 knots) must be monitored by skippers of such craft.

It is rare to experience Force 8 winds inland but gusts to this speed and higher can and do, occur particularly when there are showers, thunderstorms or cold fronts passing.

**Hills and Mountains**
Winds and weather in mountainous areas is very variable and can be treacherous. No one should venture into upland areas without adequate preparation and wearing the right clothing to allow for the sudden deteriorations that can occur. It is always important to ask knowledgeable locals about their weather vagaries.

Winds will be steered by the local hills and valleys and may bear no relation to the wind direction culled from the forecast isobars. In some places, where winds are funnelled, they can be strong but the fact that the wind is strong in one place means that it has to be light nearby. The terrain will affect the wind much more in winter than summer but any overcast day is likely to see very variable winds between one locality and another.

Cloud builds very rapidly over slopes and one important situation is the cool clear morning where the sunshine tempts the unwary out for a long walk or climb. Showers, which will be cold and can be very heavy, will often break out before lunch-time with consequent risk of exposure to those unequipped for the weather. In spring and autumn what would be rain, or even dry, on low

ground may be snow or sleet at altitude. In muggy airstreams remember the risk of hill fog.

When the airstream as a whole is stable, lens-shaped (lenticular) clouds form in waves across the wind direction and can lead to strange winds at the surface which blow against the prevailing direction. These may be accompanied by gustiness with no predictable pattern. Even those at sea need not escape because the effects are felt tens of miles downwind of the hill ridge that is causing them.

**Hoar frost** is the white frost which forms on grass etc on frosty mornings. If freezing fog is responsible for the frost then the trees will also become covered in hoar frost.

**Humidity** is the degree of wetness of the atmosphere. In lay terms it is expressed as relative humidity (RH) which can be found from a paper hygrometer such as are sold in garden centres. One must not rely too heavily on the readings of such cheap devices and the only sure way of finding the RH is to use a wet-and-dry bulb thermometer (hygrometer) which can often be bought at ironmongers or DIY stores.

The wet-and-dry bulb hygrometer has two identical thermometers with the wet bulb surrounded by a wet muslin wick. Evaporation from the wet bulb cools the latter. The dryer the surrounding air the greater the rate of evaporation and so the greater the cooling of the wet bulb compared to the air temperature registered by the dry bulb. The difference

| Height | | Pressure | Temperature |
|--------|------|----------|-------------|
| metres | feet | mb | °C |
| 15,000 | 50,000 | 120 | −56 |
| 9,000 | 30,000 | 300 | −43 |
| 7,000 | 23,000 | 400 | −30 |
| 5,000 | 16,000 | 500 | −17 |
| 3,000 | 10,000 | 700 | −4 |
| 2,000 | 6,000 | 800 | 0 |
| 1,000 | 3,000 | 900 | +8 |

between the readings can, with Hygrometric Tables supplied with the hygrometer, give the RH and the dew point.

In fog and cloud the RH is 100% as the air is saturated while over arid deserts the RH may be as low as 10%. Usually the RH is around 80% over land and 90% over the sea although the 80% figure will fall in the interior of continents and especially in winter. Only a whirling psychrometer (a wet-and-dry bulb hygrometer made to whirl in the air like a football rattle) can be accurate enough to give those at sea an adequate value for the dew point in order to predict the possible incidence of fog.

**ICAO Atmosphere** This gives the average values of temperature, pressure, density etc for reference purposes. Rough values suitable for our purposes are as the table on the next page.

**Instability** is a term which means that the general tendency of the air is to rise. Unstable airstreams need cool air over warmer surfaces and then showers and possibly thunderstorms can break out. The fact that cool air sometimes does not produce big heap clouds is due to either there being a strong inversion which limits the cloud tops or that the air is too dry. However any airstream that produces heap clouds is unstable but only up to the cloud tops. Instability is a feature of day-time weather over land but the same airstream can be unstable throughout the twenty-four hours when cool air comes south over increasing sea isotherms.

**Isotherm** A line joining places at the same temperature.

**Isothermal layers** Isothermal means 'without change of temperature'. Thus an isothermal layer is an air deck in which the air temperature remains sensibly constant with height. From our point of view they act like inversions. The tops of alto cloud layers will often be determined by isothermal layers at altitude.

**Katabatic winds** Winds that flow downhill as opposed to anabatics which flow up-hill. Katabatics begin to blow as soon as the sun leaves steep hill and mountain slopes.

**Kata fronts** are fronts over which subsiding air is leading to the erosion of cloud and the cessation of any precipitation. They are generally older fronts and are past the ana-front stage where the air in them was mainly ascending. Some kata fronts are so weak as to only consist of some thick stratocumulus cloud out of which a little drizzle, light rain or light snow falls.

**Latent heat** Whenever a substance

condenses from vapour to liquid or changes from liquid to solid a certain amount of heat has to be given out and carried away by the surroundings. This heat does not appear to be there because the temperature does not alter as the change takes place. This is why it is called 'latent' or 'hidden' heat. It is accounted for by the fact that the molecules are much more energetic in the vapour than in the liquid phase. Likewise they are more energetic in the liquid than in the solid phase. As each energetic molecule is forced to become less active it gives away its excess energy as heat. The sum total of an astronomical number of molecules doing this adds up to the latent heat released.

In the case of water, when a kilogram of water vapour condenses to cloud droplets over 200 million joules of heat are released into the surrounding atmosphere of the developing cloud. That is about the same as the heat you get from one bar of an electric fire in forty minutes.

The release of this heat slows the natural way in which air cools as it ascends (the lapse rate) but it also fuels updraughts in cumulo-nimbus clouds as the heated air rises.

As clouds will contain millions of tons of water so the amount of heat released when clouds form is immense and leads to the formation of inversions and iso-thermal layers at altitude.

The amount of heat released when a kilogram of water solidifies into ice is only a seventh of the above but it still adds up to a sizeable quantity when it occurs in the amounts inherent in weather processes.

Latent heat is a reversible process and so when ice melts it has to take in the same quantity of latent heat as it gave out when it formed. Thus melting snow cools the air and adds to the general rawness when there is a thaw. Equally to make puddles disappear after rain requires a great deal of heat to be supplied either from the sun, from warmer air (as happens when a warm front has just passed) or from the earth below the puddle – or all three.

When one considers the immense bulk of some icebergs and the amount of latent heat energy they must absorb in order to melt, it is not surprising that, having calved on the ice shelves and glaciers of the Arctic and Antarctic, they survive for a very long time and so threaten the shipping lanes.

Once they have formed, cloud banks do not disappear very rapidly because the latent heat they emitted when they formed now has to be supplied by the sun or by warming due to subsidence. If there is no tendency for the air to sink any further or it is winter with low sun or it is night the cloud layers have no real tendency to disperse and simply persist. Thus great banks of cloud for which there is no immediate reason will be seen drifting around on the cloud satellite pictures.

**Local weather** The weather can vary very considerably between one neighbouring locality and another. Factors which induce these differences include going from mainly flat to undulating terrain plus proximity to the sea or maybe a large lake or reservoir. If the locale sticks out into the path of say northerly winds then snow showers and showers generally are more prevalent than over neighbouring areas which experience shelter from that particular direction. Showers from the sea will survive only so far inland as they were receiving their moisture from the sea and have now deposited it as rain. The cloud survives but the precipitation dies out.

Quite modest hill barriers will make the windward sides more prone to precipitation than on their leeward side which will be drier. Mountain barriers that stand across the prevailing wind will provide shelter for all the terrain downwind of them. Any gaps in the chain will allow showers to feed through into the plains beyond.

Even depressions are most likely to feed through the gaps in mountain barriers. For example, depressions tend to enter the Mediterranean through the Gironde Gap and the Straits of Gibraltar.

**Local winds** Those who live within 30 or so miles of the coast will be subject to sea-breezes which tend to blow in spring and summer. The effects are to make the coastal hinterland much sunnier than further inland. They cool the afternoons of hot days but can make a part of the late morning hotter than it might otherwise be when the breeze stems an opposing wind and leads to calm or very light winds.

Nocturnal winds can become strong when the coastal plain is backed by sizeable hills or mountains and it is the right time of year. Snow-capped peaks inland from any coast spell the risk of falling winds that may locally reach gale force.

Katabatic winds also become strong when snow-capped mountains are involved and may threaten craft sailing on lakes, tarns, lochs and fjords. Anabatics are not likely to be strong but provide part of a mountain wind system in many of the valleys of the Alps and similar mountain chains.

Supercells or large areas of summer thunderstorms will produce local winds that bear no relation to the general wind flow for the area.

**Millibar (mb)** This is the standard pressure unit used by meteorologists and is a pressure of 100 pascals (Pa). A thousand mb is a bar. So the bar is a pressure of 100,000Pa and is a useful unit as it is close to standard atmospheric pressure.

The isobars on many TV weather maps are spaced at 4mb intervals but professional meteorologists tend to draw them at 2mb intervals while in the continent of Europe they may be drawn at 5mb intervals. Newspaper charts covering large areas such as the North Atlantic will often be drawn with isobars at 8mb intervals. These give the general positions of lows and highs but a great amount of the essential detail is lost.

**Monsoons** are regional winds which are akin to very extensive and long-lasting sea and land breezes. They blow with

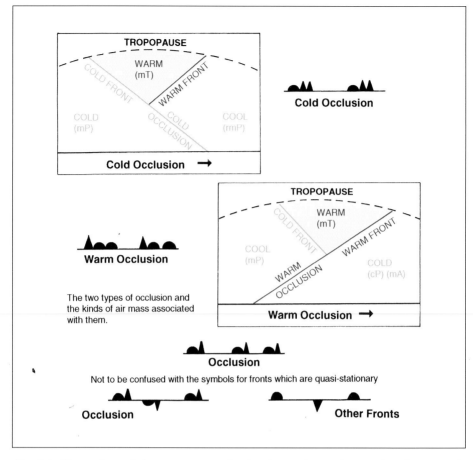

The two types of occlusion and the kinds of air mass associated with them.

Fig O1 Illustration of the two types of occlusion.

great consistency and regularity in sub-tropical areas such as the Indian sub-continent. Nearer the equator the differences between land and sea temperatures are not sufficient to produce monsoons and in temperate latitudes they tend to get lost in the general run of the westerlies.

**Occluded fronts** are fronts where the cold front has overtaken the warm front of a depression and lifted the warm air off the ground. The ultimate goal of weather processes is to raise air warmed by the Earth's surface and replace it with cooler air. Thus occlusion is the natural end to the life of a depression and spells its decline into eventual oblivion. As Fig O1 shows there are technically two kinds of occlusion but for the layman the difference is largely immaterial. In both the warm frontal characteristics come first to be followed, without a break, by the showers and then easing light rain of a cold front.

Most of the fronts which cross the British Isles and move into Europe are occluded. The cloud build-up ahead of occluded fronts will often be quite rapid and recognising the passage of an occlusion means that the air to come will be fog-free polar air rather than a warm sector with its risk of poor visibility.

**Radar** can detect rain, hail etc and networks of weather radars now cover areas such as Great Britain and large parts of the United States. These see where showers and larger areas of rainfall are occurring in real time and enable immediate warnings to be given to major sporting events as well as more general bad weather warnings when major storms are beginning to develop.

It must be remembered that individual shower clouds have a limited life and may only last half an hour, however as one cloud dies so another often takes over. So the radar that sees a group of showers at this moment will not be seeing the same clouds in an hour's time.

Radar cannot however see 'dry' clouds but sometimes heap clouds can be detected because of reflections from birds who are using the updraughts for soaring.

**Radiosonde** Every day at noon and midnight GMT balloons are released all over the world to sample the temperature, pressure and humidity of the upper atmosphere. The balloons carry a reflector which enables them to be followed by radar and so the strength and direction of the upper winds ascertained.

This information, together with the surface observations, is fed to the computers in order that they can work twice a day from real values. In this way the forecast charts start from a solid base and any anomalies that may have crept in can be ironed out.

**Rainbow** When the sun is not too high in the sky and rain is falling, rainbows may be seen. They are best seen late in the day when scattered showers are passing and you have to have the sun behind you.

Everyone sees their own rainbow, even though we agree as to where it is and what the colours are. This follows because the colours are due to refraction through spherical raindrops. As shown in Fig R1 the light from the sun is bent and reflected within the individual drops back to the observer. There is a particular angle of refraction where the light becomes very concentrated and this is different for each colour. Thus we see the same colour at the same angle to the eye and the result is a bow.

If the light is very bright it reflects again within the drop and this light now emerging at different angles to the primary bow forms a secondary bow. Very occasionally a third bow may be seen but personally in a lifetime's observing I have never seen the third bow.

Rainbows do not have any great forecasting value other than to prove that a cloud, that maybe passed without wetting you, has now decided to rain and therefore perhaps there are more showers to come.

**Saturation** Take a polythene bag and put a few drops of water in it. Seal it off and if you can still see water drops in the bag after a while then you know that the air

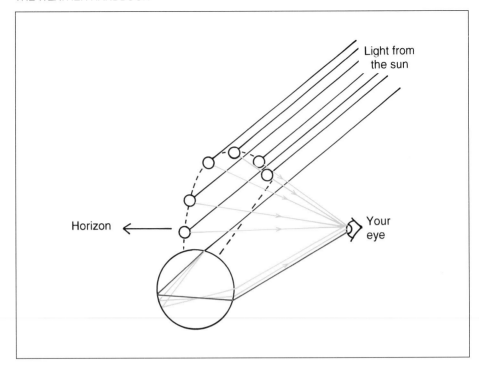

*Fig R1  How you see a rainbow because each colour is refracted in the raindrops at a slightly different angle and reflected inside them.*

in the bag is saturated with water vapour.

This state of saturation comes about because as many water molecules are condensing onto the water drops as are evaporating from them. If not the water drops would evaporate away and disappear and they obviously have not done so.

Cool the bag down and more water will appear. The air in the bag is in a state of supersaturation. Warm it up and eventually the water will disappear as all the molecules will now be invisible vapour. We now say the state of the air is 'dry'.

In the atmosphere supersaturation is naturally the way that clouds form and all kinds of precipitation occur. Clouds die out when they become dry because of warming processes in the atmosphere.

**Specific heat** The sea is a vast reservoir for heat because water will store more heat in unit mass than any other natural substance. This ability to accept or emit heat is called specific heat, and the latter is defined as the quantity of heat required to raise the temperature of a kilogram of water by a degree C. It takes no less than 3,900 joules of heat to raise a kilogram of the sea by a degree and equally this same quantity of heat has to be taken away when each kg of sea cools down. This is eight times the heat required to say, heat up a kg of iron by a degree and accounts for the high bills when we have to heat water by electricity or gas. The S.H. of earth etc is a very variable quantity but is much less than water.

**Sublimation** When white granular hoar frost appears on grass, twigs etc the water vapour does not go through a liquid phase but condenses straight out as ice. This process is called sublimation.

**Sunshine** The sun sends us light and heat as if it were a furnace at 6,000°C. How much of each form of radiation we get can be gauged from Fig S1 and it shows the reason why we see so well. The eye has developed to take in the brightest band of wavelengths that the sun emits. The vast mass of the sun's radiation comes to us as invisible infra-red (IR) wavelengths which produce heating in absorbent materials. Rough matt black surfaces absorb best and so heat up while shiny surfaces reflect most of the radiation and so remain cool. You only

have to put a sheet of kitchen foil under the grill for a few minutes to find, often to your surprise, that you can handle the foil without burning yourself. Do not however try the same trick with a well-used baking pan.

Only a percentage of the sun's total power reaches the surface being reflected by clouds etc and then some more is reflected by the surface. This is

*Fig S1 The curve that tells us how much of each wavelength we get from the sun. There is obviously much less total UV than IR and less visible than either of them, but the visible light is radiated at maximum intensity and our eyes see the middle of the spectrum best which is why we see so well.*

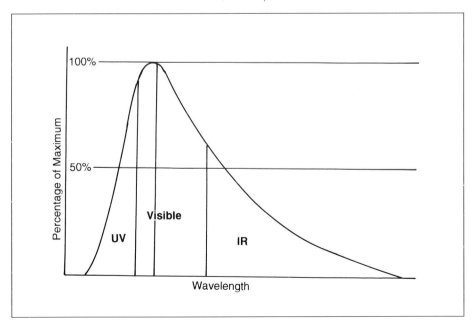

especially so of stretches of water, while a snow surface reflects a very high proportion of the sun's rays. Some more of the incoming total is absorbed by $CO_2$ and water vapour in the atmosphere. Only 43% of the sun's radiation is absorbed by the Earth and much of this is by the land masses.

The ultraviolet component of the sun's radiation can be harmful to the exposed body. Interaction with $O_2$ oxygen molecules in the high atmosphere leads to the formation of $O_3$ i.e. ozone, which is a highly effective selective absorbent of the harmful incoming UV. It is the loss of some of this protective shield which is giving rise to worry about harmful UV effects on this and coming generations.

For sunscreening purposes the UV band has been split into UVA and UVB bands of which the B are the most harmful but it has been found that A are also harmful. To guard against skin cancers it is recommended that a broad spectrum screening preparation is used. If practicable UV-proof sunglasses should also be worn.

**Tornadoes** are usually found in the right rear of intense thunderstorms. They are violent whirls of air a few hundred feet in diameter with winds exceeding 200 knots near the centre. They are most prevalent and damaging in the southern United States where their effects are so amazing as to defy explanation. In Britain and Europe generally, tornadoes are quite prevalent but as they do not often lead to death or injury they do not get much publicity. The damage they do to caravans, tents, fences, sheds etc as well as the roofs of houses is by no means negligible. There is a special tornado warning service in the areas of the US most prone to them.

**Tropical revolving storms** are very intense circular disturbances 100–150 miles across with winds of hurricane force and tracks that are directed away from the equator of either hemisphere. They are called Hurricanes in the north Atlantic and most of the Pacific. They are Cyclones in the Arabian Sea, the Bay of Bengal and south India while in the China Sea they are Typhoons and in North-west Australia they are called Willy-willies.

The normal track of Atlantic hurricanes is shown in Fig T1 and forecasting them has been transformed by satellite pictures. The U.S. geostationary satellite *GOES East* which is over the Amazon basin can keep constant watch for the emergence of hurricanes and track them from then on allowing evacuation of threatened areas to be undertaken in an orderly fashion. Hurricanes feed on warm sea which supplies heat and moisture. Thus when a hurricane comes ashore it loses its impetus very rapidly.

Atlantic Europe sees old hurricanes in September and October when they have recurved across the Atlantic. They can be rejuvenated by an injection of cold polar air and become quite vicious leading to autumn storms.

**Turbulence** When the wind blows over rough terrain it becomes full of eddies. The sum-total of these eddies is called

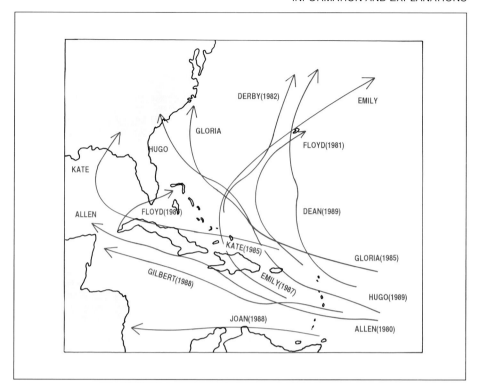

turbulence and the height to which turbulence extends depends on the wind speed and the degree of instability in the air. The normal height to which turbulence extends is 2,000 feet but this entails a wind of about Force 4 over land when there is a normal degree of instability.

The turbulence mixes the air-deck it affects and makes it all come to a much more equable state of wind speed, temperature and humidity. The night-time inversion layers prevent mixing and lead to a lowering of turbulence with a consequent drop in wind speed. The wind speed above the inversion increases to compensate.

Over the sea the depth mixed by turbulence may be almost halved compared to its value over land under

*Fig T1 Some of the hurricanes that occurred between 1980 and 1989 and their names. The names start with A for the first identified and go on through the alphabet. So the devastating hurricane Hugo was the eighth hurricane in the 1989 season.*

the same circumstances. It is turbulence that makes the wind variable with a wind vane constantly swinging about around the mean wind direction.

**Veering** Winds are said to veer when they shift direction clockwise. The opposite is backing, i.e. the wind shifts in a direction back against the direction the sun moves.

# Index

# Bibliography

The following are a list of titles which will either give deeper information or give a different slant.

*Air Rider's Weather*. Alan Watts. Published by A & C Black. Wind and weather as it applies to para and hang gliders and balloons.

*Elementary Meteorology*. HMSO.
An authoritative explanation of modern met. written in terms the layman can understand. For ashore and afloat.

*Hazards in Mountaineering*. Paulke and Dumler. Published by Kay and Ward.
Gives excellent advice for all those who venture into the uplands.

*Heavy Weather Sailing*. K. Adlard Coles. Revised by Peter Bruce. Published by Adlard Coles Nautical.
The acknowledged textbook on extreme conditions for yachtsmen.

*Instant Weather Forecasting*. Alan Watts. Published by Adlard Coles Nautical.
Twenty-four colour pictures of the sky and what inferences can be drawn from them.

*Instant Wind Forecasting*. Alan Watts. Published by Adlard Coles Nautical, and Sheridan House, N.Y.
Suggestions on what the wind may do for small craft sailors.

*Meteorology for Mariners*. HMSO.
A book written with merchant navy officers in mind.

*The World of Weather*. Brian Cosgrove. Published by Swan Hill Press
A comprehensive, large-format book on the world's weather. Superb colour photography and diagrams throughout.